D1055503

THE
INCIDENTAL
STEWARD

THE
INCIDENTAL
STEWARD

Reflections on Citizen Science

AKIKO BUSCH

ILLUSTRATIONS BY
DEBBY COTTER KASPARI

Yale

UNIVERSITY PRESS

New Haven & London

Published with assistance from *Furthermore: a program of the J. M. Kaplan Fund* and with assistance from the foundation established in memory of Philip Hamilton MacMillan of the Class of 1894, Yale College.

Chapter 3, "Weeds on the River," was first published in a slightly different form in *TriQuarterly* 133.

Names have occasionally been changed in order to protect the privacy of the people quoted.

Yale University Press books may be purchased in quantity for educational, business, or promotional use. For information, please e-mail sales.press@yale.edu (US office) or sales@yaleup.co.uk (UK office).

Designed by Lindsey Voskowsky.
Set in Adobe Garamond type by Integrated Publishing Solutions.
Printed in the United States of America.

ISBN 978-0-300-17879-1

A catalogue record for this book is available from the Library of Congress and the British Library.

This paper meets the requirements of ANSI/NISO Z39.48–1992 (Permanence of Paper).

10 9 8 7 6 5 4 3 2 1

For my father, Noel F. Busch

The future enters into us,
in order to transform itself in us,
long before it happens.

Rainer Maria Rilke

CONTENTS

ACKNOWLEDGMENTS

When guest editor Donna Seaman invited me to contribute to the journal *TriQuarterly,* she put me on the road to documenting these excursions. I am greatly indebted to her. No less so to everyone I spoke with in these projects—scientists, educators, volunteers, activists—all of whom were unfailingly generous with their time, knowledge, and expertise, not to mention unfathomably patient with my questions. My thanks goes as well to my agent, Albert LaFarge, for knowing just where this book belonged; and to my editor, Jean Thomson Black, for her insight, constant encouragement, and the absolute clarity with which she always responded to my doubts and questions. Her assistant, Sara Hoover, was always there, instantly helpful every step along the way. Laura Jones Dooley has been thoughtful, perceptive, and precise, the copy editor every writer hopes for. William Schlesinger, president of the Cary Institute of Ecosystem Studies, graciously invited me to a residency at the institute, and I am grateful to him, as well as to all the scientists there, for reminding me about the value of uncertainty; that questions are often more important than the answers; that the facts of the natural world are overwhelming and contradictory; and that there are rarely clear answers. That was, and remains, a good place to begin. It is an honor to have Debby Kaspari's elegant and evocative drawings included here. Lisa Dellwo's editorial counsel and surgical finesse have been of immeasurable value, and I am grateful as well to Anne Kreamer and Sian Hunter for sound advice and support at crucial moments. Steve Stanne—I thank you for sharing your well-spring of knowledge and for your exceptional generosity. Allan and Julie Shope's Great Wall of Eco-Heroes was a resource and an inspiration. Furthermore, a program of the J. M. Kaplan Foundation, was generous in its support of this book, and I am indebted to Ann Birckmeyer and Joan Davidson for their confidence in this project. A grant from the New York Foundation for the Arts also provided me with time to develop this material, and that assistance has been invaluable. Last, my thanks, always, to Brian.

Acknowledgments

1

Introduction

We do not understand ourselves
yet and descend farther from heaven's
air if we forget how much the natural
world means to us.

—Edward O. Wilson

Mohonk Mountain is only thirty miles from where I live in the Hudson Valley, but the ascent always makes it seem farther. On the November day I take the drive, mist hugs the ridge and drifts over the valley. The trees have long since shed their foliage, but for those few leaves lingering on some oaks, and those will hang on for most of the winter. Lichen and moss appear to soften the stone ledges they blanket, and a few pitch pines grab onto the rock. The mountain laurel keeps its green into the winter months, but the real tenacity is in the stone itself. Shawangunk conglomerate is an enduring quartz composite that sparkles when you look at it up close but from a distance is a pearly white. The serrated cliffs of white rock look to have been edged out of the earth's crust in a great, sweeping uplift, which is why Mohonk is often called a "sky island."

Mohonk Mountain House, along with the preserve around it, was founded in 1869 when Albert Smiley purchased several hundred acres ninety miles north of New York City. Believing that a knowledge of and engagement with the natural world enriched one's life—spiritually, intellectually, physically—Albert, along with his twin brother, Alfred, continued to buy adjacent land on the ridge, committed to preserving the land both as a resort and as a nature sanctuary at a time in our country's history when these purposes were not mutually exclusive. In pursuit of that goal, the Smiley brothers established a weather station in 1896. Temperature and rainfall have been recorded every day since then, providing scientists with one of the longest-running weather records in the nation's history.

Daniel Smiley, nephew of Albert and Alfred, transformed this basic system of documentation into a broader and more comprehensive catalog of information. Recognized now as one of the country's most renowned naturalists, he continued to take temperature readings twice daily from the late 1930s until his death in 1989.

But along with data about temperature and rainfall, Daniel Smiley's observations extended to whatever occurrences of animal and plant life he might happen to encounter on the nearly nine thousand acres of the preserve and mountain house property, whether sunspot cycles, spring bird arrivals, the first bloom of plants, the breeding habits of salamanders, the nesting preferences of peregrine falcons, the fungus on the dogwood trees, the appearance of gypsy moths, or the disappearance of blueberries. He recorded all these observations on small, three-by-five-inch note cards, and today that archive of cards makes for a comprehensive record of phenology, the study of seasonal phenomena. In an age of specialization, when scientific research tends toward particularized, narrow fields of study, Smiley's generalist approach may seem an anachronism, but in fact, that very inclusiveness is what gives the data their value.

Daniel Smiley confessed to two attributes in the family's genes: the perpetual need to record what was observed and the conviction that nothing could ever be thrown away—document and keep, a simple but sound basis for research. Today all of these records, more than a century's worth of natural history data, are cataloged in the Daniel Smiley Research Center, which was founded in 1980. After Smiley's death, his longtime assistant Paul Huth continued his work. Today a young environmental geologist, John Thompson, is bringing the research center into the electronic age. Although the same equipment is used now as then—official Weather Service thermometers, a brass gauge issued by the US Weather Bureau to measure rainfall, and an iron bar bolted to the conglomerate in 1899 to measure the level of water in the glacial Mohonk Lake—Thompson is charged with creating a digital map for records going back to 1925. With a continuity that is consistent with the way the study of nature is practiced here, he is the same age as Huth was when *he* began as a young botanist to work with Dan Smiley in plant identification. The composition of the quartz is not the only thing that endures up here.

Late November 2011 seems like a good time to take the drive up to the research center. Earlier, record summer rains from Hurricane Irene and tropical storm Lee had saturated the Hudson Valley, and months later, the river was still running high and brown from an overload of runoff silt from tributaries. At the end of October, back-to-back freak snowstorms had left us with more than seventeen inches of snow, trees halved by the weight of snow on their leaves, and a five-day power outage. Now, in late November, it has been unseasonably warm for a week; the forsythia on the side of the road is in flower, and the goldfish in my neighbors' pond are

3

Introduction

spawning. The long-term climate records will possibly decipher these seasonal aberrations, and indeed, an hour or so later, as we are sitting around a table in the research center, Thompson tells me that the 8.21 inches of rain recorded in one day during Hurricane Irene was the most ever logged in a single day; that the wettest August on record was followed by the wettest September on record; that these were followed by the October with the greatest amount of recorded snow; that the average temperature has risen more than two degrees in the last 116 years; and that warmer temperatures have added ten days to the region's growing season.

Of greater value to me than the actual data is the template of attentiveness that the research center offers. One hundred sixteen years of continuous weather data and about a hundred thousand records on note cards represent a means of understanding the minutiae of the natural world. Meteorologists, climatologists, botanists, and geologists have long considered the Mohonk archives a model of dependability and constancy. Though not a scientist, I have come with the hope of learning about this mode of precise observation. It has a place, I am certain, in the way we all can be witness to a landscape in transition. Deliberate and detailed, it also recognizes ambiguity. In a profile of Daniel Smiley published after his death, he is described as "one who seeks objective answers and models, but never quite accept[s] evidence as a final truth."[1]

At Mohonk's Smiley Research Center, paying attention is a twofold enterprise. Smiley was methodical, precise, regular in his observations about everything from rainfall measurements to bird arrival dates to the management of gypsy moths. At the same time, he put a high value on the incidental. Although the serendipitous find was never a replacement for calculated research, unexpected sightings were always recorded. A note card about the sighting of a porcupine reads in full:

> 11 Sept '30. A. K. Smiley, Jr. saw one on Woodland Drive at Lake Shore Path. Mohonk Lake, NY. He was followed and headed almost straight across country till he left the Forest Drive at the sharp turn below the spring. He was not apparently foraging but headed somewhere. This was about dusk. He passed within about two feet of Ki [Keith Smiley] paying no attention to him. As far as I know this is the first record here. I have heard that they are found in the Clove.

This note card was filed away in the card catalog with dozens of others having to do with porcupines. Its significance would come later.

This is a way of taking in the world that begs the question: When does a fact acquire meaning? When it is observed? Later, when it converges with other facts? Or later still, when all of those facts collaborate into some simple truth, if they ever manage to do that? When do data become knowledge? It occurs to me there is no simple answer to this question. Maybe the most you can say is that a sense of place is likely to be derived through some sense of accumulation, through some vast collection of sightings, observations, impressions, records.

Smiley's impulse to take notes is something many of us seem to share, whether it's the legion of gardeners who make it a habit to jot down what's blooming when or lake communities whose newspapers report when winter ice begins breaking up. But the long-term data sets especially valued by scientists studying global climate change can come from the notes of venerated naturalists. The young Franklin Roosevelt kept a field notebook of birds he observed, notebooks put to use a century later by a twenty-year-old researcher to extend her studies correlating the earlier arrival of certain migratory birds with climate records.[2] Likewise, scientists from Boston University studying how the flowering times of certain species of plants in Concord, Massachusetts, respond to a warming climate turned to records made by Henry David Thoreau in the mid-nineteenth century.[3] Such information is of value to scientists analyzing transitions in the natural world, whether bird migrations, amphibian habitats, the arrival of nonnative species, or changing weather patterns. Long after science has become professionalized and the province of PhDs, American naturalists are in a position to gather and provide reliable data for research. And if what they practice is called "citizen science," it is because their processes, equipment, and motives are a product of contemporary times.

Dan Smiley's manner of observation looks to both convergence and divergence. This is not simply a way of watching the natural world but a way of thinking about it. The first, convergence, is the knowledge about the interdependence of natural events; when it comes to matters of climate, plants, and animal species, one thing often leads naturally to another. To observe the natural world is to observe a complex web of relationships. The second, divergence, has more to do with unexpected happenings, the unpredictable occurrences and serendipity that nature continually tosses our way. It is impossible to take in the events of the natural world without being alert to both kinds of events.

Maybe another way of saying this is that there are two ways of paying attention: when you know what you are looking for, and when you don't.

My own efforts to pay attention had begun on a May afternoon in 2007 when a research assistant from the New York Department of Environmental Conservation pulled his truck up in front of our house to ask our permission to search out a radio-tagged bat that had signaled it was roosting in the woods behind our house. I followed him on his trek up the mountain. In the months and years that followed, this outing led to others—to afternoons on the Hudson River or along its tributaries, to streambeds and vernal pools, to visits with scientists and educators who let me tag along on their fieldwork. Some of these endeavors followed strict protocols with data sheets and precise checklists. But others looked to a more casual agenda. Sometimes attentiveness was a solitary endeavor, sometimes it included another human being, and sometimes it was a collective effort. Occasionally it required dividing up the landscape into quadrants and imposing a grid over it. Periodically it was about seeing, counting, measuring what was there, making calculations, determining what had been gained and what had been lost. Sometimes it was about observing failure. Other times it was simply about bearing witness to the unexpected.

All of which is to say, it was a matter of following those elusive procedures that are a part of caring for anything else in life that matters. And if these outings followed different rhythms and protocols, such variations are reflected in these pages; the chapters here do not offer a primer on the protocols of citizen science. They are not drawn from a traditional field notebook full of facts and figures or from the precise entries noted on data sheets. Instead, they are looser dispatches and include the marginalia, scribbled asides, side roads, and digressions that one so often follows on such unpredictable excursions.

Beyond the shared inclination that drives people to record events of the natural world or to enter into volunteer collaborations with scientists, I was motivated as well to make some sense out of a changing landscape. I had grown up in the Hudson Valley during the 1960s, when it was a farming region. There were few coyotes and bald eagles then. I have no memory of robins wintering here. Mile-a-minute vine had not advanced along the creek beds of Dutchess County, nor did banks of purple loosestrife proliferate along the wetlands in August. And although zebra mussels had not yet found their way to the Hudson River, it was then a toxic stew. DDT in the pesticides streaming into the river was inevitably consumed by the river's bass, herring, and perch, and when eagles ingested the contaminated fish, their eggshells thinned, diminishing the chances for full incubation. When I look back to those early years, I also remember the grandeur of towering elms lining an avenue near our house before the advance of Dutch elm disease.

In 1987, when I was in my early thirties, I moved back to the Hudson Valley with my husband. After twelve years of living in cities, I had come to know it was the etch of branches against a winter sky, the smoky hue of the woodlands in November, or the way the color leaches back into the earth in March that all spelled home to me. In returning, I had found a place in a state of change. The white-tailed deer and the occasional black bear we find in the yard are there because there are too many of us and we are disturbing their native habitat. A maternal colony of Indiana bats has taken up residence in a black locust tree up the mountain; there is speculation that they have been relocating here from southern states because they require a cool climate. And along with the zebra mussels, invasive water chestnuts and black boss have helped to reconfigure the ecology of the Hudson River just to the west. Yet at the same time, the bald eagle population has been restored, and it is not uncommon to see a dozen eagles soaring over the ice floes. The coyotes in the woods behind our house are another recent arrival, and forest recovery and wildlife management programs have allowed as well for the restored populations of wild turkeys, beavers, fishers. And last March, a neighbor came upon a moose in the middle of the road that winds up the mountain behind our house.

More unusual species have started to turn up as well, and those sightings have taken on a different, less logical rhythm. Some have an almost hallucinatory quality, as on the morning I rounded a corner of the road leading into our valley and was surprised to see the town supervisor, standing at the side of the road with a lasso in his hand and a confused expression on his face. Twenty feet away was a large, ungainly bird, at least five feet tall with pale blue feathers, an emu, I later learned, that had escaped from a local breeder. Or the summer day when I arrived home to find four peacocks sitting on the roof of our house, as though exiles from some forgotten bit of folklore. Escapees from one of the farms in my neighborhood that have taken to cultivating exotic breeds, they spent the morning fanning their palace of feathers and chortling on the peak of the roof. That afternoon, they settled into high branches of the blue spruces in front of the house. All day we listened to their cackle and cry. The next morning they were gone. And not long ago, a friend told me about driving home on a June evening after a dinner out and finding a small, green African parrot in the road, its shimmering iridescence bringing an unexpected tropical shine to the summer dusk. He picked it up and took it home, fed it, gave it water, and a day later managed to locate its owner.

If there is something surreal in these exotic animals, nonnative plants have a more established presence here. Many of these—the Norway maple, garlic mustard, thickets of common reed that proliferate on the river's tidal marshes, sprays of multiflora bushes that seem to fringe every meadow—have long been naturalized to the point that many of us assume they are native species. Others, though, have not been absorbed so easily. Mile-a-minute vine can race along a creek bank at the rate of six inches a day, while the giant hogweed, with its baroque, fifteen-foot-high stems and four-foot leaves, reflects new and extravagant standards of intrusion. The persistent speed of the first and the lavish scale of the second speak to the extremes with which nonnative plants are taking up residence in the local landscape, the rate of their arrival undiminished.

It is hard not to become engaged in, even mesmerized by, these changing habitats and animal and plant populations. Some have disappeared forever, while others are arriving. The natural world is a place that is defined, of course, by continual change and evolution; it is a place defined by unexpected sightings and unpredictable events. Or, in the words of Loren Eiseley, "There is nothing very 'normal' about nature."[4] Still, it is the rate of change and its sometimes absurd character that is so striking now; the series of unexpected alliances and bizarre appearances often suggests a segment of the Discovery Channel directed by Federico Fellini. Warming temperatures in the Hudson Valley have led to more extremely hot summer days (above ninety degrees Fahrenheit) and a decrease in the number of cold winter days (below thirty-two degrees Fahrenheit). Continuous records maintained by the water treatment facility in Poughkeepsie indicate that the Hudson River has warmed 0.945 degrees Celsius since 1946, with the past seventeen years being many of the warmest.[5] The timetable of spring has been re-adjusted in recent decades: apple trees bloom eight days earlier than they did in the 1960s, grapes six days earlier, and lilacs four days earlier. Bird arrival dates have shifted, too, the killdeer, American woodcock, tree swallow, and green heron all coming days and some-times weeks earlier. When such arrivals do not coincide with simi-lar shifts in the insect species they feed on, entire bird populations can be threatened.[6]

Sometimes I like to imagine that in the woods behind the house an unlikely gathering of all these creatures is taking place, the parrots, peacocks, and emus along with the cardinals, woodcocks, and green herons convening to address the reconstruction of their habitats, some fantastic convergence of species that speaks to the eerie détente with which we all seem to coexist now. Perhaps swing-ing overhead from a length of bright yellow twine like some merry decoration is a triangular purple basket, a purple prism trap for the emerald ash borer, a tiny, metallic green beetle that is eating its way through American ash trees. The gathering is a little festival of dislocation. At such times, I can't help but wonder if this is not a landscape in transition so much as one in a state of derangement.

People talk about having a sense of place, but what does this ballet of extinctions and arrivals do to that?

In the previous century, absurdism was the province of art and literature. Salvador Dalí's melting clock faces, Eugene Ionesco's il-logical interchanges, and Samuel Beckett's muted dialogues all de-livered us to the tangential associations between thought and action, to the elasticity of time, to the unfathomable nature of modern human experience. Today, anyone intrigued by the disassembly of logic can simply go outdoors. Although it may be less clear who authored the scene, who is the director, and who is the audience, who, in fact, dreamed this all up, the same disjointed reasoning and disconnected sequences will be fully evident. It is February as I write this. Migratory glass eels have appeared in the Hudson River weeks earlier than their presence has ever been recorded; Japanese apricot trees and camellias are flowering in New York City; the snow-drops and daffodils in my own garden have sprouted. The rivers and shorelines, forests and fields are now the stage on which the incomprehensible circumstances of modern life are most clearly played out. Dutchess County in 2012 is a landscape beset by con-tradictions, a place where there is more forested area than there was a century ago but less wilderness; a place where we look to rustic traditions, value homegrown food, and savor seasonal rituals, all the while besieged by emerald ash borers, mile-a-minute vine, and runaway emus; a place with increasingly erratic weather patterns, rising temperatures, rainfall surges, and record flooding.

My family had moved to an old, gray, shingled farmhouse sur-rounded by woods and meadow in New York's Hudson Valley in 1962. One day, not long after we had settled in, I walked the boundaries of our thirty acres with my father. He wanted to post the land so that local hunters would not shoot deer on it. And so we spent the afternoon tacking little fabric signs to trees marking the boundaries of the property, posting the signs every thirty or forty feet on maples and oaks, hickories, beech and ash trees. "NO TRESPASSING," each sign read in large black letters, with an expla-nation of the conditions of trespass in smaller letters below, along with my father's name. He then also pointed out to me two markers for the property: whereas two of the four corners of our land were clearly indicated by landmarks at the road front, the two corners in the wooded area were marked by large gray, flat rocks.

Decades later now, I find myself often thinking of that walk with my father. My husband and I never had enough land to bother posting, but even if we had, I know that laying claim to place is a different exercise for us, having less to do with tacking signs to trees and putting stone markers into place. It is more difficult to define. The best I can do is to say that it has mostly to do with watching, with paying attention. And all of us—my neighbor who came upon the moose, my friend who found the parrot, the supervisor charged with catching the emu—may find ourselves coming to such attention, to some collective sense of observation suffused with curiosity, apprehension, protectiveness, some concurrence of emotional human responses that seems particular to these times. It only makes sense.

How we have become estranged from place is well documented; it has become part of our cultural profile. The Census Bureau reports that one in six Americans moves each year. And certainly where we live, local newspapers have all but vanished. As the postal service continues to be besieged by fiscal troubles, small post offices across rural America close, another landmark of community gone. Regional architecture exists largely as nostalgic remnants; the mom-and-pop store on the corner has long been replaced by a national franchise with a generic metallic mansard roof that could just as well be in Wisconsin or California. Communications technology, for all its miracles of connectivity, can further dilute our sense of place; because cell phones, text messages, email, and Skype all facilitate our route away from where we happen to be, it is easy for our allegiance to *where we are now* to fray. And global positioning systems readjust our spatial frame of reference. Whereas a conventional map situates the user within given landmarks and boundaries, with the user shifting position as passage is made across the printed page, global positioning systems place the viewer at the center of the screen. The relocation diminishes our sense of geographical context, undermining our ability to form the cognitive maps of place that traditional printed roadmaps help us to build. "Break your GPS, and you may find yourself lost," says one psychologist studying how our sense of space is affected by this technology.[7]

Ecopsychology is a branch of psychology that suggests there is a connection between the health of the individual and the health of the natural world; that the psyche needs the textures, rhythms, and cycles of the natural world to remain intact; and that when the relationship between the human mind and nature has broken down, a pathology results. In an effort to define this emerging field of research, Daniel B. Smith wrote in the *New York Times* that "just as Freud believed that neuroses were the consequences of dismissing our deep-rooted sexual and aggressive instincts, ecopsychologists believe that grief, despair, and anxiety are the consequences of dismissing equally deep-rooted ecological instincts." It is a view of human behavior also closely tied to biologist, naturalist, and author Edward O. Wilson's theory of biophilia, which asserts that our connection to natural systems is intrinsic to who we are; that it has an evolutionary component; and that "a sense of genetic unity, kinship, and deep history are among the values that bond us to the living environment. They are survival mechanisms for ourselves and our species."[8]

Are the psychic disorders we suffer the consequence of living in a world that has become ecologically deranged? In my mind, it remains an open question whether a sense of displacement from the natural world brings us greater anxieties or whether our other existing stresses cause us to distance ourselves from nature and cause this larger sense of being adrift. I would suspect that both of these are true and that, in fact, displacement is likely to travel this infinite loop; the greater our disquiet, the more we tend to remove ourselves from nature; the more we become so removed, the more deeply our anxieties take root.

But all of that said, or perhaps as a direct result of it, the search for some sense of belonging seems to befall many of us, whether it is in watching, naming, counting, documenting, or otherwise noting the natural events unfolding around us. The *Hudson River Almanac* is an online natural history journal that gathers weekly observations on or about the river. Contributors may be scientists and researchers, but they are just as likely to be ordinary homeowners or schoolkids who live on or near the river and are paying attention. Observations, annotated by the Hudson River Mile where they occur, range from the comings and goings of the great blue herons, egrets, cormorants, and terns to the sudden appearance of a giant swallowtail butterfly or gray seal. They capture what the tide brings in and what it takes out.

An entry notes, "Autumn has been displaying its immense range of river conditions, from three-foot rollers caught between wind and tide to a day like today when the Catskills' remnant colors are mirrored on the Hudson's surface. Spotted sandpipers dipped in flight over the water, and along the shore an adult bald eagle was almost camouflaged in a near-white sycamore." Another states, "Special moments can occur when the stealth of the observer matches that of the observed. As I stood upwind and absolutely still, an eight-point buck stepped out of the tree line not fifty feet away. I hardly breathed and my eyes burned from not blinking. The white-tailed deer walked slowly across a narrow opening and dissolved into the shadows of another tree line." And another: "There was a quarter-inch of ice on the puddles this morning. A hulking coyote with an enviable thick pelt trotted to the top of the landfill and monitored my walk. I spotted small flocks of red-winged blackbirds and a score of robins with their 'companions,' cedar waxwings. Bluebirds were scattered across the leeward hillside, bluer than the cloudless sky."[9]

Dan Smiley might have been pleased, I think. Like his notations, all of these come from volunteers simply being present. And documenting. And keeping. Like his, these notes offer a kind of spare poetry, a record of what is there with an unembellished lyricism. And like his observations, these record both the "convergency" and "divergency" of natural events, taking in both expected seasonal events as well as unanticipated occurrences. As such, they compile a broad database for the ecology of the Hudson Valley. The almanac entries and the Smiley note cards may both also reflect an impulse that is timeless and universal: both also are modern iterations of the distant directive uttered by the seventeenth-century Japanese poet Basho: "Go to the pine if you want to learn about the pine, or to the bamboo if you want to learn about the bamboo."

In our conversation up at the Smiley Research Center, John Thompson had mused on the idea of replacing the daily visits to the weather station with automated equipment. Up to eighteen parameters are measured each day, he tells me, among them rainfall, fog, snow, freezing rain, thunder, misting, haze, hail, ice pellets, and frost. "You might get a lot more data," he admits. "You could get readings every fifteen seconds if you wanted to. But without walking up there, you would miss what you didn't see. You might not see the solar halo. Or you might miss the first phoebe of spring. You just don't know what's there unless you see it with your own eyes."

Still, seeing something with your own eyes may not always be enough. When does a fact acquire meaning? An observed fact is not necessarily a known fact, and I consider another, different daily record of events. Not long after the Deepwater Horizon oil rig exploded in the Gulf of Mexico on April 20, 2010, a live camera feed was installed five thousand feet below the surface of the water. Until the leak was capped on July 15, the footage of the spill was something visible to us all, the plume of black spewing oil continually pictured at the lower right-hand corner of the TV screen. But though we all watched exactly the same image for weeks and weeks, the visual documentation of the spill before our eyes, we were unable to arrive at a consensus about what it was. Initially, it was considered to be a thousand barrels a day, then five thousand barrels a day, then forty thousand, then sixty thousand. This was a billowing cloud of poison that no one—whether marine ecologist, engineer, fisherman, oil driller, or news analyst—knew how to measure. You can see something and not know what it is.

Whether a bluebird bluer than a cloudless sky or a submerged plume of oil, both speak to the imagination that is required to see what is right before our eyes. And if that imagination is apparent in the *Hudson River Almanac,* it has surfaced elsewhere as well. In my outings, I encountered it in the efforts of a 9/11 widow who found solace and an abiding sense of continuity in pulling water chestnuts out of the Hudson River. It was reflected in the experiences of a laid-off IBM manager who had found a use for her administrative skills in documenting vernal pools and again in the routine labors of a Catholic priest who goes out on his tractor to mow the invasive vine that is choking the meadows and woods next to his church. And it occurs to me now that all of them may be seeking objective answers and models, though never quite accepting evidence as a final truth. Maybe that's one way the imagination manages to compose itself.

It seems safe to say that at a time when our sense of place can be so frayed, these efforts all suggest ways to restitch it; when so many things conspire to help us forget where we live, these endeavors can reestablish our sense of place; they remind us that place matters; they reposition us, they root us. As one teenage girl helping to count the migrating eels in a Hudson River tributary told me, "It's *fun*. And it's extra credit. And it's something I'm doing for my community. And then I'm just gonna learn more about eels than you ever knew was there"—surely words and work that echo Dan Smiley's conviction that "nature has aesthetic, philosophical, and spiritual values for men."

All the encounters on these pages are specific to the Hudson Valley, its plant and animal life, the river's tributaries and wetlands and the fields and woodlands surrounding these. But the care and attention I found among the people participating is *not* specific to place. Similar enterprises are taking place elsewhere around the country. After the BP oil spill, legions of volunteers, dubbed "turtle people," in Alabama coastal towns came together to collect the eggs laid by gigantic sea turtles, pack them into coolers, and arrange their transport to Florida beaches that were not drenched with oil. Working with the Florida Fish and Wildlife Conservation Commission and the National Marine Fisheries Service of the National Oceanic and Atmospheric Administration, ordinary citizens implemented the egg evacuation effort. In New Hampshire, a retired schoolteacher and veteran maple farmer uses elementary schoolkids to help her collect and research the sap from trees diseased by warming temperatures. In California, volunteers in the California Roadkill Observation System record animals killed by cars. Their photographs, GPS coordinates, and species information provide a comprehensive catalog for roadkill—which is then used to analyze the environmental impact of roads; and how roads can be better built, maintained, and marked. And Operation Migration is the name of an effort to restore the population of endangered whooping cranes to the eastern United States by guiding juveniles in their flight training; working with biologists, "craniacs," dressed in white hooded suits, tutor the cranes in reclaiming their flight paths. In these cases, and so many more, volunteers who have an inclination to pay attention and can use a smart phone or manage a GPS unit can be effective agents for environmental stewardship.

The term "citizen science" first appeared as the title of a 1995 book by British sociologist Alan Irwin, though Irwin's content was more engaged with science, technology, and social policy. That same year, Rick Bonney, then director of education at the Cornell Lab of Ornithology, used the term in a grant proposal to the National Science Foundation, recalling: "I came up with it while looking out my office window as an alternative to 'amateur scientist.' I had no idea how broadly it would subsequently be used."[10] Certainly, the term seemed apt for the lab's long history of volunteer initiatives—ranging from its Nest Record Card Program (1965) and Project Feederwatch (1987) to the broad variety of programs it manages today (see Appendix). The lab states on its website that the term "citizen science" has been "used to describe a range of ideas, from a philosophy of public engagement in scientific discourse to the work of scientists driven by a social conscience. In North America, citizen science typically refers to research collaborations between scientists and volunteers, particularly (but not exclusively) to expand opportunities for scientific data collection and to provide access to scientific information for community members. As a working definition, we offer the following: projects in which volunteers partner with scientists to answer real-world questions."

Others take a broader view. Judith Enck is the Region Two administrator of the Environmental Protection Agency and works with a somewhat more inclusive definition. "The EPA can't be everywhere at once," she says. "We need issues brought to our attention. People are the eyes and ears of the community. Our commitment to environmental justice makes citizen science vitally important to us." Enck states that the agency depends on citizen science to help identify the source of public health issues, whether air pollution, water pollution, or something else entirely. As she puts it, the community is the first to know the problem, whether it is hazardous air pollutants that are causing asthma or wastewater discharge from a combined sewage overflow that may be contaminating local water sources. Enck cites a video on YouTube documenting a wall of untreated sewage moving down the Gowanus Canal in New York City after a heavy rainstorm. These filmmakers, she says, "are true citizen scientists," who, she suggests, can be using "any tool focusing on scientific measurement or portrayal."[11]

Bonney now tends to use the acronym "PPSR," which stands for Public Participation in Scientific Research, an umbrella phrase that covers a variety of research models. As he stated in a 2009 paper, these models involve the public to varying degrees. Bonney divided the projects discussed in the paper into three categories: "contributory projects, which are generally designed by scientists and for which members of the public primarily contribute data; collaborative projects, which are generally designed by scientists and for which members of the public contribute data but also may help to refine project design, analyze data, or disseminate findings; and co-created projects, which are designed by scientists and members of the public working together and for which at least some of the public participants are actively involved in most or all steps of the scientific process."[12]

My own preference here is to use the term "citizen science" and to leave its definition intentionally broad. I may not be sure that any and all public participation in scientific research is citizen science, but I do believe that the enterprise is in such a nascent phase as to warrant, for the moment at least, a comprehensive view. Partnerships among volunteers, scientists, and regulatory agencies may be loosely drawn affiliations—sometimes in documentation and research, other times in remediation and restoration. As Bonney states, "There are so many different models out there. I don't really care what they call it, just so long as it is done well. And works."[13]

Whether they are working locally, regionally, or nationally, citizen scientists are continuing a cherished American tradition. Amateur naturalists, and how they pay attention to the rhythms and disturbances of nature, have always had a place in our country's cultural life. Their attentions are embedded in our heritage. The indigenous peoples lived here with an abiding respect and reverence for the world around them. The Shawnee warrior Tecumseh battled deeded land rights for most of his life, and his rending question resonates in the environmental movement today: "Sell a country! Why not sell the air, the great sea, as well as the earth? Did not the Great Spirit make them for all the use of his children?"

When our forebears arrived here centuries ago, their search for a new life and new form of governance extended to an appreciation for and curiosity about the nature of the new continent. Thomas Jefferson was known for his scientific inquiry, the plants and trees imported to Monticello, his unwavering interest in flora and fauna, the mountains and streams of his home state, Virginia. As a farmer, he studied crop rotations, seed development, and soil cultivation. John Muir studied geology and botany but never received a degree. Ralph Waldo Emerson attended divinity school. John Burroughs was a teacher, a farmer, and an employee of the Treasury Department. John James Audubon went to military school and was a teacher and a taxidermist before painting his dazzling catalog of American birds in their natural habitats. And Dan Smiley graduated from Haverford College with a degree in engineering rather than botany or biology. What informed their work was not a formal education in the sciences but a belief in the convergence of science and humanism, a commitment to stewardship, and the conviction that full engagement with the natural world enriches the human experience.

Citizen scientists today uphold this legacy. Like their predecessors, their work tends to reflect an ethos proposed by the feminist theologian Sallie McFague. She advises us that, as housemates who share the planet's resources not only with other living organisms but with its future tenants, we must abide by three rules: "Take only your share, clean up after yourselves, and keep the house in good repair for future occupants. We do not own the house. It is loaned to us for our lifetime with the proviso that we obey the above rules so that the house can feed, shelter, nurture and delight those who move in after us."[14] Yet several factors also distinguish today's stewards from those generations preceding them. We may be compelled by a new sense of urgency; as Al Gore has observed, watching the evening news is something like a nature hike through the Book of Revelation. Whether it is climate change, a burgeoning human population of seven billion, habitat loss, extinctions in the world's oceans, or thawing permafrost, we live in a time when catastrophes of our own devising are visited regularly on the natural world; whether it is toxic red sludge in the Danube, deforestation in the Amazon, or radiation in Fukushima, the systematic cycle of assault is just as regular and certain as those of season and climate. It would not be going too far to say that an impending sense of crisis drives a new and different sense of engagement in ordinary people.

Add to that the host of new technologies such as global positioning systems, digital photography, Internet databases, and interactive websites that can assist in collecting and sharing reliable information. The records of volunteers can be put to good use. And we are, of course, experiencing a period of unprecedented advance in new digital tools. The next generation of sensors, for example, is likely to be smaller, cheaper, more networked; these will allow air and water monitoring to be a far broader effort, extending from the individual and community levels to municipal, state, and federal.[15] Elsewhere, research and development look to new trends in Do-It-Yourself enterprises and open-source systems. Consider the Public Laboratory for Open Technology and Science, an international initiative that offers tool kits for inexpensive aerial mapping: the balloon, line, work gloves, and accessories for rigging a digital camera enable ordinary people to create aerial maps with which to monitor and record spills, leaks, and other environmental adversities that can best be documented overhead.

Likewise, the economics of using volunteers to gather a broad spectrum of data points can have value for the scientific community whose funding sources are ever more strained. The desired information is widely visible. Climate change may affect species in four ways: the range in habitat; changes in size, appearance, or behaviors; genetic frequencies; and phenology.[16] The flowering of plants, the spring arrival of songbirds, and the population of butterflies are all phenomena that come easily to public attention. Broad-based volunteer groups may also have access to private lands that might otherwise be difficult to access through traditional research venues. A 2010 article in the *Annual Review of Ecology and Systematics* went so far as to state, "A large suite of applied and basic ecological processes occur at geographic scales beyond the reach of ordinary research methods. Citizen science is perhaps the only practical way to achieve the geographic reach required to document ecological patterns and address ecological questions at scales relevant to species range shifts, patterns of migration, spread of infectious disease, broad-scale population trends, and impacts of environmental processes like landscape and climate change."[17]

A final quality shared by many naturalists today may be their more collaborative approach. For their predecessors, the consideration of nature was often a solitary endeavor; they made their observations and documented their discoveries on solo sojourns that were often an effort to escape the frantic pace of society—Henry Thoreau's cabin by the pond, John Muir's walk to the gulf, Aldo Leopold's shack by the river. Now, just as often, naturalism tends to be a collective enterprise, a community of like-minded citizens. When the snowy owls migrated from the Arctic to the lower forty-eight states in the winter of 2012 in unexpectedly high numbers, their sightings were reported on eBird, a site run by the Cornell Lab of Ornithology and the National Audubon Society, enabling the vigilant birders to share information and generate broader public interest and knowledge. In the words of Edward O. Wilson, "The stewardship of environment is a domain on the far side of metaphysics where all reflective persons can surely find common ground."[18]

Participants engaged in all of these endeavors may just want to get outdoors. Or they may be motivated by a general interest in conservation. Perhaps they are monitoring conditions in their community that may have public health consequences. They may even be looking for a more serious engagement with specific issues and their potential impact on education, research, and policy. But while spending time in the natural world often brings a sense of restoration, that restoration becomes much larger if the result has some societal value. And if it furthers the partnership between scientific research and land use. Surveying vernal pools may be a regenerative way of witnessing spring, but the experience grows in meaning when its results have a tangible effect on how local planners consider the value of open space. Taking inventory of the water celery beds in the Hudson is a good excuse to spend the morning kayaking on the river, but when the data reflect measurable facts about water quality, the connection to the river goes a little deeper. Every year, my friend Doug Reed participates in "A Day in the Life of the Hudson River," during which environmental educators team up with schoolkids to collect sediment samples, test water quality, identify fish, measure salinity levels, and otherwise learn to gather and share data about the estuary. Participation doubles, triples each year, he says. "They keep coming back. They're hooked. That's what happens after a student takes the first step into the river with a net."

All of these efforts, too, may be steps toward what William Schlesinger, president of the Cary Institute of Ecosystem Studies in Millbrook, New York, calls "translational ecology," a response to the increasing urgency for ecologists to find ways to deliver their information to the public and to policy makers. "Translational medicine," he observes, helps physicians teach patients about new research and practices in health care and, in so doing, brings medical research into drug development and treatment; a similar partnership might ensure that ordinary citizens learn "the implications of scientific discoveries and understand their impact on alternative ecological diagnoses." And connecting ecology to stakeholders, he suggests, "should enhance the understanding and application of ecological concepts, ensuring that scientific rigor is brought to bear on the world's many environmental challenges."[19]

When stakeholders are connected to ecology, the bond can run deep. Whether it is stepping into the river with a net, pulling up a length of invasive vine, or some other effort in experiential activism, it's not much of a stretch to suggest that rediscovering a sense of order in the natural world is a way to find our bearings. Possibly it is a way to tap into what ecopsychologists call the "ecological unconscious." Finding the water celery is a way of picking up the loose thread, locating a vernal pool a means of answering to some dry spell. Anytime one looks to the natural world, a larger lesson plan appears. Purple loosestrife is an introduced plant species, and our management of it can speak to those ways in which we fold the unfamiliar into our lives.

My own outings here may have started as part of some effort to calculate the distance between the Hudson Valley I knew as a child and the one I live in today. Maybe tallying the eels entering a Hudson River tributary or looking for the herring that come up the river to spawn in May or counting the eagles for the midwinter census count could supply some of those numbers that help to gauge that distance.

And since then I've wondered, too, if these excursions were some effort to continue that walk in the woods with my father so many years ago. From time to time over the years, I have revisited those woods. More than once I have cyberstalked them, finding that Google Earth is capable of returning me to those trees and meadows. From my desk, with my cursor, miles and years away, I sometimes do my best to follow those familiar paths. But it is not easy to do, the trails overgrown with weeds, the thickets too tangled, the apple trees gone. And the screen on my laptop does not have the resolution to tell me whether the woods of maple, beech, and birch are giving way to oak and hickory, as has been predicted. Nor is it clear whether the populations of chickadees and purple finches are on the decline while those of the indigo bunting and cardinal are increasing.[20]

But sometimes my trip is accompanied by a different soundtrack. Thewildernessdowntown.com is an interactive film created by director Chris Milk that invites visitors to visit their childhood homes. After they have typed in the address, Google Maps, Street View, and music by the group Arcade Fire collaborate to take the visitor there as the lyrics sing out, "When the lights cut out, I was left standing in the wilderness downtown." A film captures a young man running down a street, his footsteps pounding, a constellation of birds exploding in the sky. The camera pans the bygone streets and neighborhood, then pulls back, and the landscape spins.

"Write a postcard of advice to the younger you that lived there then," the website suggests.

I remember the hundred thousand or so observations that Daniel Smiley wrote over a lifetime on index cards and the thousands of notations that have been made over the decades in the Hudson River Almanac. And I wonder at this innate and persistent desire to keep ourselves connected to the places we love, whether it is on an interactive music video, in a digital almanac, or on three-by-five-inch green note cards. All of them recognize some essential protocol about documenting and keeping. All are good blueprints for the postcard I imagine I might write. "I miss you, I miss who all of us were, I miss our lives there together," I might say. I might find myself more anxious than that. Perhaps I would say, "It is all so long ago. But where are the apple trees? " Yet that wouldn't begin to say it all either.

So the chapters that follow came to work as a letter to the place I once lived, more epistolary than the meticulous field notebooks kept by the scientists I encountered, citizen or otherwise. And the excursions here came to be some series of outings parallel to that walk in the woods I took with my father forty years ago. But because "trespass" has come to have a broader meaning now than it did then, I have tried to arrange the words on these pages as some revised version of those small fabric signs, some equivalent way to mark out, to know, to claim the place where I live now.

2

Bats in the Locust Tree

From time to time during the summer of 2007 I found myself trekking up the mountain behind our house. It would be dusk, the hour the bats leave their roosts, and it was the bats I was after. Pregnant and nursing bats cluster to rear their young, their shared body heat essential to the well-being of their progeny, and just such a maternal colony of Indiana bats had taken up residence in the woods on the mountain. Though added to the endangered species list in 1969, Indiana bats had begun to stabilize in number by 2000, and more recently, their range in northern states, including New York, has increased. While nothing here points definitively to climate change, the shift in range could possibly be caused by changes in forest growth. During winter, the bats hibernate in groups of tens of thousands in a variety of known mines and caves, but their scattered summer populations are less well known. In an effort to understand their population shift, biologists from the Division of Fish, Wildlife and Marine Resources of the New York State Department of Environmental Conservation had been tracking the bats' summer range, collecting data about seasonal movements and habitat needs; species survival could depend on this knowledge. And species survival has wide-reaching implications. A lieutenant in the nocturnal ecosystem militia, one bat can devour thousands of insects in a night.

About sixty bats had been counted at the roost up the mountain, but whether that number was constant remained in question. Local residents who conduct exit counts as the bats leave the roost in early evening are useful, because they can save the crews from the Department of Environmental Conservation from making the hour-and-a-half drive down from Albany for a twenty-minute count. Although it's a citizen science project to which I would like to contribute, I know there's another reason that I make this occasional trip up to the dying black locust tree where the bats have taken shelter: its withering branches speak to the way the dislocated will find a way to be accommodated by even the frailest system.

The first time I'd made the hike in early May, it had been with

Jeff Carson, a twenty-two-year-old intern who had volunteered his time on the tracking study. Jeff worked with Al Hicks, a biologist in the New York State Department of Environmental Conservation's Bureau of Wildlife in Albany. A couple of weeks earlier, as some thirty thousand Indiana bats had emerged from hibernation in a mine in Rosendale to the west across the Hudson River, twenty had been live-trapped. Hick's team had recorded their weight, size, gender, reproductive conditions, and time of capture and fed them mealworms, crickets, and water, keeping them in a quiet, warm place before releasing them with tiny radio transmitters attached with a surgical adhesive to their backs. With a battery that lasts about twenty days—about as long as the adhesive lasts—the transmitters allowed the bats' path to be traced. A plane equipped with GPS mapping software tracked their initial flight direction, after which ground crews tracked them to their summer roosts, where their numbers would be monitored.

Jeff had volunteered his time on the study, and cruising our road with an antenna and receiver, he had picked up a signal coming from someplace in these woods. It was partly out of sheer curiosity that I had asked to tag along. But there was something else at work, too, which had more to do with the innate human desire to witness good luck firsthand: when good fortune comes your way, you want to see what it looks like.

It was still early spring, and winter marked the side of the mountain with water seeping from beneath the rock ledges. The gullies formed by melting snow had been cut into even deeper incisions by a driving rainstorm just a week or so earlier. Now the side of the mountain seemed almost serrated, and in places the ground gave way to mud. Even at this altitude there were unexpected pools.

But here and there were a smattering of wild asters, shoots of garlic weed, and fringes of groundcover just beginning to sprout from under the dead leaves. What leaves were out were tinted that acid green of early spring, but there weren't many of them. Ancient stone walls crossed the wooded hillside, marked also by the occasional old stone foundation. A slope with a shifting degree of incline, its topography is uneven, an easy rise that gradually gives way to a stand of pines, then a clear-cut with grasses flattened by spring rain and crossed with deer trails. As the hill grew steep again, all of these variations caused the radio signal to bounce around, making for an uncertain tracking route.

Time and again, we thought we might be there. The bats roost in trees that have deep crevices and hollows or enough loose bark for them to burrow under, and the forest of oak, elms, maples, birch, hickory, and locust offered a variety of possible havens. A thicket of dead elms alongside a wetland was surely suitable, but the signal remained constant. A few minutes later, we came across a stately old shagbark hickory. "That would be the perfect place," Jeff said, but it was with more resignation than hope, because she wasn't there either. So we kept hiking up the mountain in that manner of tracking that requires advancing, then doubling back, on a zigzagging, tentative path that is familiar to anyone who has ever looked hard for anything. Although old hunting trails and deer paths offered alternate routes, we were following a more elusive path of sound. "This is as tough as it's ever been," he said, as much to himself as to me. The French writer and philosopher Gaston Bachelard used the phrase "desire paths" to identify those improvisational routes that people create when they are trying to go from one place to another most efficiently. Although landscape architects and planners devise walkways, then delineate them with paving materials, people nonetheless often find more direct and intuitive ways to get where they are going. Here in the woods, whether formed by water run-off, deer, or wild turkey, every path seemed to be a desire path.

But when we were some hundred feet below the ridge line, the signal picked up, and its steady, loud chirping took us to the old black locust tree, seventy or maybe seventy-five feet tall, with cracks and fissures in its deeply furrowed bark that made it an accommodating roost. A pamphlet from the US Fish and Wildlife Division of the Department of Environmental Conservation describes the Indiana bat as having fur that is "dull grayish chestnut rather than bronze, with the basal portion of the hairs of the back dull lead colored. This bat's underparts are pinkish to cinnamon, and its hind feet small and delicate." I saw nothing like this, but when we listened very carefully, we could hear their hum of gentle chattering. Bats use sound to situate themselves; their capacity for echolocation is what guides them, what elucidates their sense of space and substance, what locates their prey. But the sound, a series of tiny clicks ten or twenty, forty or even two hundred times a second, is at a frequency beyond most human hearing; it is exactly that high frequency with its short wavelengths that produces the detailed echoes the bats rely on to learn the shape of their universe. We live in an acoustical empire we will never know, our threshold of hearing situated somewhere between the low, infrasonic rumbling of elephants and the high-pitched ultrasonic rasping of tenrec shrews that rub their quills together to communicate. I knew that the subtle whispering I could make out was but a fraction of the sound sensation the bats were creating and, listening to the soft murmuring, felt I was standing at the gate to a remote kingdom of sound that could only be imagined.

"It's a whole colony," Jeff said. He measured the diameter of the tree at breast height: sixteen inches. He recorded the tree's condition: dying. He noted the roost type: bark. He noted the coverage of bark on the tree: over 80 percent. He observed the canopy cover: intermediate. He took its GPS coordinates so that when the crew came later to take an exit count, they'd know how to find the tree. He noted all of this on a data sheet, that one-page document meant to record the essential information of a site that is also a call for order, an attempt to catalog.

The last thing Jeff did was encircle the tree with orange ribbon to flag it. I knew that its purpose was strictly functional, but still, at the end of our search for the bats, there couldn't help but be something festive, even a bit celebratory about tying that bright ribbon around the tree trunk. Along with the lace of the early leaves and the tracery of the branches, its bit of gay trimming seemed to herald the bats' moment of homesteading and our own little carnival of discovery.

And then for a few minutes, we just stood and listened again to those distant syllables. The locust tree was approximately thirty miles from the cave in Rosendale. The project overview informed me that bats have been known to travel over four hundred miles to find a roost, though most travel less than fifty. I was heartened to know that this colony did not have to travel so far to find a secure roost. I asked Jeff why they had chosen a roost so close to the ridge line. "I honestly don't know why," he said. "Maybe there's a cavity in the tree. . . . Maybe it was born here." Bats can live up to thirty years. Who could say exactly what gives this old black locust near the ridge of the mountain such allure.

Afterward we hiked back down the mountain. Although it was only early May, the hedge of wild blackberry brambles that had grown in where the woods meet our lawn was already dense, and we had to hack our way through it. But it suddenly made sense to me to have this demarcation between our little cultivated patch of grass and garden and the more unpredictable habitats up the mountain. In Chinese folklore a bat represents the five great happinesses: health, wealth, long life, good luck, and tranquillity. While I lay no claim to this excess of happiness, there was something hopeful in knowing that this old black locust at the top of the mountain now served as host to a maternal colony of bats. It was more than just the satisfaction of finding the thing you were looking for. The low chattering resonating from this tree was oddly reassuring, perhaps only because its distant symphony spoke to the way all of us are driven to find our place as the conditions and circumstances around us shift.

It was only about five o'clock, several hours before the bats would leave the tree for the night. Jeff was headed north to where his family lives, so he left the exit count to a crew who'd come a day or so later to get an exact tally of the bats as they left the colony at dusk. But rather than follow the haphazard route Jeff and I had followed through the brush and up the mountain from my house, it'd be easier for them to start at the top of the mountain and come down the some hundred feet to the flagged tree. So now he needed to get permission from the landowners to come through their woods.

Before we got in the car, though, he tried to give me an antler we had stumbled across on our hike up the mountain. It had been lying under leaves at the foot of a small oak tree where a young buck had rubbed it off earlier in the year, and I had missed it completely. It hadn't evaded Jeff's notice, though, and he had put it in his jacket pocket, a souvenir of the afternoon. Now he wanted to leave it with me, partly I think because instinct instructed him to leave it near its native environs, and partly because an afternoon in the woods always brings out one's innate generosity. I had met Jeff only by chance a few hours earlier, yet I had the feeling that the impulse to give comes naturally to him. When I had invited myself along on his trek up the mountain, he hadn't hesitated, and he was one of those people who seem willing to share whatever he knows. A gracious landscape makes you want to be the same, and I couldn't persuade Jeff to take the shining bit of curved white antler with him. Ecopsychologists maintain that the anxiety and stress endemic to modern living is often a consequence of a relationship with the natural world that has fallen out of balance. Their field of study is the link between mind and nature, human perception and environment. Now I wondered if this was an example of that balance, for a moment, being righted.

We drove to the top of the mountain, then wound our way down through side roads until we found the small red house perched on the ridge line that we had earlier just been able to glimpse through the trees. No one was home, but the next-door neighbors were out on their deck. Their lawn, recently seeded, was sprouting. They were happy to hear about the bat habitat. The trees, still bare, offered us an expansive view across the valley. The man told us he was from Yonkers and his wife from Queens. They had just relocated a couple of years ago, having bought the house from a builder who'd built it on spec. "Only problem with the house is he put those two windows in the wrong place," he told us and pointed to two windows on the south side of the house; the wide view to the valley is to the west. But he gestured to the stream running through the adjacent woods and told us how he listens to it when he lies in bed at night. "I love the sound of it," he said.

His wife pointed to some lawn furniture that got banged up by what they thought was a bear they sometimes think they hear, crashing through the woods. "What else could it have been?" she asked. Then she asked Jeff where they might best put up a bat house or maybe even a house for purple martins. Birds have a preference for edge habitats, he told them—where the forest meets the fields or where the marsh abuts the pine stand. Or in this case, that boundary between their cultivated lot and the rough, wooded hillside that lies between their house and mine. By now it's the far edge of the afternoon, too, and I looked down the mountain in the direction of my house on the other side of these woods and wondered if it's some similar impulse that brings all of us to these edges. Even more I wondered at how indistinct and blurred they were becoming, and how much longer such edge habitats would be available to any of us.

Months later I learned that the bats occupy another edge as well. Earlier that winter, wildlife biologists had learned of a new threat to the bat population. An explorer in upstate New York had observed bats flying erratically outside the caves in midwinter, often during daylight hours. Subsequent research had found clusters of dying bats inside the caves and out, their disease signaled by a ring of whitish fungus across their nose that is sometimes compared to an eerie, improvisational death mask. Bats afflicted with the fungus had a reduced content of body fat, preventing them from surviving the cold; they were starving to death. Emaciated female bats have a hormonal imbalance that prevents reproduction. Studies conducted since 2007 have indicated that bats with the fungus are awakened more easily from hibernation and for longer periods of time. The increase in metabolism causes them to burn through fat quickly; then they either starve to death or fly out of the cave in search for food, where they tend to freeze to death or get eaten by predators.

One of the few certainties surrounding the disease is that it can be transmitted easily through the bat population. The fungus grows in cold temperatures and can spread quickly through the tight clusters in which Indiana bats hibernate; by now it has catastrophically reduced the bat population across the northeastern United States, by some estimates claiming more than five and a half million of them. Although there is not yet any known way to treat the disease, a first step has been to close caves where the bats hibernate to public access, and caves in eleven states that served as bats' hibernacula are now empty. The 2007 emergence study had set out to analyze the emerging bat population in the northeastern United States, and learning about their summer habitats was a way of ensuring species security. But what began as a study of population patterns and migration morphed into one of possible extinction.

From time to time over the following weeks I found my way back up the mountain. The number of bats in a maternal colony can range from a dozen to several hundred, and when I had spoken to Al Hicks on the phone, he had told me that the exit crew had counted some sixty bats roosting in the black locust. Any count I could give them would help determine how fixed such colonies are. But it was rare that I saw the bats on these treks. The canopy had filled out, and if the woods resonated with the songs of cardinals, thrushes, and sparrows, the quiet murmuring of the bats was harder to discern. I learned that they typically have six to eight roosts and rotate between them in an irregular pattern. Sometimes I saw them flitting from one tree to another, but without an antenna and receiver, it was difficult to pinpoint the roost or make any exact count.

And the truth was that I was of little use to the project, a fact I acknowledged with frustration. The best I could do was to watch the trees where the bats continued to skirt my line of vision. It was an uncertain science, but it had its own rewards. If I was unable to determine the roosting tree, I at least might encounter a bush of wild blackberries near an old foundation. Or, one evening, a perfect half moon slipping in behind the leaves. But that's the thing about spending an hour in the woods—if you don't find what you're looking for, you're sure to find something else—and I recall Dan Smiley's belief about convergence and divergence.

I also remember Jeff's precise checklist of observations, but the page I am confronting now is more like the one described by John Muir, who once wrote,

> When a page is written over but once it may be easily read; but if it is written over and over with characters of every size and style, it soon becomes unreadable, although not a single confused meaningless mark or thought may occur among all the written characters to mar its perfection. Our limited powers are similarly perplexed and overtaxed in reading the inexhaustible pages of nature, for they are written over and over uncountable times, written in characters of every size and color, sentences composed of sentences, every part of

a character a sentence. There is not a fragment in all nature, for every relative fragment of one thing is a full harmonious unit in itself. All together form the one grand palimpsest of the world.[1]

My own notes probably lie someplace between Jeff's efficient data sheets and Muir's inexhaustible pages of nature, erratic notations that make it clear I do not yet have the attentiveness for such observation; it requires a kind of looking and listening to which I am unaccustomed. Even when I know what it is I am after, my watchfulness is capricious and irregular, evoking a game my mother would sometimes play with my sister and me when we were growing up. We would gather with our friends, and she would show us a grouping of objects on a black lacquer tray, perhaps a bracelet, a teacup, a fan, a toothbrush, a coin, a cookie, a leaf, a dish, a ruler, a random assortment that might or might not commonly be found on a kitchen tray. She would give us ten or fifteen seconds to look at them and assign them to memory. And then, after the tray was removed, we would be asked to list what we had just seen, and I might write "bracelet," "fan," or "leaf," but it was rare that I remembered even three quarters of the items.

Now, after a walk in the woods, I often find myself trying to summon up what has just been presented to me on that black lacquer tray, to register for another tutorial to correct the astonishing ability human beings have to dismiss what they have seen with their own eyes only a moment earlier. I am reminded again that so much of the nature of human experience has to do with what we see and what we remember. It is likely, though, that I suffer, as so many of us do, from what is called "inattention blindness," a term coined by cognitive psychologist to describe the oblivion so often induced by new technologies. Focused on our smart phones, text messages, and music downloads, we can become indifferent to what is happening around us, and it is what allows us to step off the sidewalk and into the path of oncoming cars and buses. But I would suspect that such visual apathy is not necessarily the result of our mobile gadgets. It comes too easily to all of us, all the time. Humans, by nature, see what they want and expect; our senses devise their own systems of discrimination. And our vision is perpetually obscured by hope, fear, anticipation, expectation, or any of the assorted emotions of the moment. The call on my cell phone may be distracting, but I distract myself just as easily. Animals have no experience of inattention blindness; their sense of place is unwavering, their survival instincts causing them to stay attuned to what is happening around them.

Black bears live in these woods, as do bobcats, and sometimes at night, I can hear the cry of coyotes. The deer and the wild turkeys that thrive here make themselves more visible. And now there are the bats to consider. Considering them may be the most I'll do. As an incidental steward, I've come to accept that I may never observe them closely. Such is the allegiance it is possible to form toward those phenomena you may never see or touch or hold. Maybe it is always in the nature of the search to find that the character of what you were looking for has shifted in the process. I had thought it was a dull grayish chestnut bat with cinnamon underparts and delicate hind feet. Instead it was a sound sifted through the summer evening, a glimmer of wing in the trees, a notion of the haven that can be offered by even the frailest system.

In trying to imagine how our small community can sustain the species relocation occurring in all manners around us, I know now that it will be useful to consider not only the bear crashing into the lawn furniture or the coyote wandering from its den but also this colony of bats that has found its way to a dying black locust tree. Such are the more subtle routes to endurance that tend to escape our notice. As things around us change as quickly as they do, it occurs to me that if any grace or knowledge is to be found in our watchfulness, it may come in the way we learn to honor what we have not seen and do not yet know. And to respect what we do not yet love.

3

Weeds on the River

The water chestnut is what is known as "an introduced species," but on a quiet backstretch of the Hudson River on a July morning, such phrasing seemed especially curious. I had always thought of an introduction as some kind of courteous exchange, an agreement on some kind of mutual civility. In the language of conservation biology, however, the term refers to non-indigenous organisms that have either accidentally or deliberately been brought to a new location, often with an adverse effect on native plant and animal life. Out on the river, it was clear that no gracious understanding of any kind had come to pass. Rather, the water chestnuts simply emerged here as they do every summer, their leaves forming a canopy over the surface of the water, preventing light from getting into the water, reducing oxygen, inhibiting photosynthesis, and in the process making it nearly impossible for the aquatic life beneath to survive.

Ralph Waldo Emerson suggested that a weed was nothing but a plant whose virtues had not yet been discovered, but as befits a time when our transactions with the natural world seem to have gone awry, it only makes sense that the water chestnut, *Trapa natans,* has followed the reverse course. Once valued for its culinary and medicinal properties, it was introduced to this country in the late nineteenth century as an exotic ornamental plant. Now, it is nothing more than a water weed that chokes our rivers, and with seeds that can attach to everything from bird feathers and fur to boats and cars, it has a reach across the waterways of the Northeast that is swift and irreversible. Which is to say, it spreads like most other kinds of trouble—whenever, wherever, and however mercilessly it can.

Yet looking at the plant, who would think it capable of such malevolence? Who could have dreamed up such a plant that at once floats and is rooted? Certainly the configuration of its parts are so particular one could only imagine it was designed for some specific purpose, though it would be hard to guess precisely what. Its rosette composed of softly triangular leaves floats on the surface of the water, while its barbed, black seedpod anchors the plant in the riverbed; often called a devil's head, the pod, with its sharp, pointed protrusions, is treacherous when stepped on. Clearly, this is a plant with a double life, one above the surface of the water, the other below. It is at once concealed and exposed, submerged and in the open air, its innate beauty and brutality tied together by nothing more than a frail stem and feathery leaves tinted watery brown and pink. Alternately, the plant has been called "death flower," "water rose," and "water caltrop" after the metal burs intended to pierce the hooves of horses in warfare.

Each season now, they come and they go. Like other agents of dysfunction, they observe their own rhythms, their own private sense of order, specific to time, place, and the conditions of the air and water around them. Annuals, water chestnuts grow best where the river is slow and shallow, and this particular patch of the Hudson River just above a spit of land is a backwater that seems made to cultivate them. Their custom is to germinate sometime in late May to the middle of June. Later in the summer, as the salt front from the Atlantic Ocean moves upriver, the salinity will kill help to them, and by the end of September, the leaves, roots, and stems will all have decomposed. By then, the sharp devil's head is just the husk for its seed, which drops down to the riverbed to winter in the mud and germinate the following summer.

When I had gone down to the river in early June, there was no sight of them. But it was late afternoon, and the tide was high, and the angle of the summer sun was such that the river seemed without color, just a wide, glistening sheet of light. There is no beach on this stretch of river, but when I climbed over the rocks to put my feet in the water, I saw that it was brown and murky. Hudson River water is rarely clear, its turbidity a result of sediment, soil erosion, plankton, salinity, and decomposing organic matter. But the water chestnuts operate with their own botanical craft, making themselves visible only at low tide, and already, I knew, they had been carpeting the surface of the Fishkill Creek just south of here. It was pointless to imagine we would be spared the water chestnuts this summer.

Besides, invasive though they may be, the water chestnuts generate mixed feelings. A friend who is a marine ecologist told me she views them as the aquatic equivalent of canaries in the mine. She believes that their annual appearance offers its own reassurance: their sudden absence could indicate that things are even more wrong than they already are. Her words are a reminder of how our index of disaster has been revised—and of all the different ways things can go wrong. That they are here at all brings distress to the river's aquatic life; but were they to vanish, that could signal even greater disaster. And it occurs to me then that we have entered upon some new abnormality in the landscape of calamity that we must consider how things are going wrong when they have not gone wrong.

And sure enough, a week and a half later, I saw they had surfaced, a radiant meadow of assassins in full bloom across that wide plane where the water meets the air. For the first week or so after emerging, their blanket across this stretch of river was incomplete, patchy rather than the continuous, opaque surface I remembered from the year before. But in a matter of a days, there were more of them, and then still more, and as they continued to fill in over the weeks that followed, it was impossible not to wonder if they were creating a habitat of their own, this city of stars floating at river's edge. In late June, when it is spawning season, I could watch the carp jumping, flashing, arcing in and out of the surface. Tolerant of water with low oxygen levels, the carp are after the macroinvertebrates that thrive on the water chestnuts or in the sediment on the bottom of the river. The young carp can also take advantage of the shelter they find in the water chestnut beds from waves and currents, as well as from such predators as largemouth and striped bass.

Clearly there was some kind of life going on here. Elsewhere, the great blue herons use the carpet of water chestnuts as a feeding ground. And citing big bluet damselfly adults, marsh lady beetles, waterlily leaf beetles, wolf spiders, and water fleas, a report prepared by the local conservation group Hudsonia remarks that "the water-chestnut-community must be a restaurant for small to medium sized predators which can cope with dense beds, water, and soft mud."[1] Surely there is a lesson about those times when calamity becomes its own reassuring habitat; those occasions both on the river and off in which a constant state of distress becomes so familiar, sometimes so deeply comforting and comfortable, as to cultivate its own small, sustaining ecosystem.

By early July, the water chestnuts had become dense, and what began as a thin, almost translucent film of leaves across the surface of the water had become a thick, unremitting carpet. Well removed from the current of the midchannel, this tranquil stretch of river offers a place to swim, but if it is to be accommodating to swimmers, some of the vegetation will have to be culled. It's a job with its own enticements: being in a cool river on a hot summer day has a universal appeal. But it's not just that a river remains a symbol of restoration for all of us, some vital artery streaming through our emotional landscape. At a time when a sense of displacement so often characterizes our transactions with the natural world, something in the act of clearing a river choked with weeds may also speak to our own collective need for order. All of which is how my friend Nancy and I came to find ourselves weeding the river.

It was in the high eighties, the water temperature only ten degrees cooler, not much relief on a blistering July day. This was a summer river, which is to say, the water at the surface was tepid, warm, but the farther one delved into the water column, the cooler it became. Because the lower Hudson River is an estuary, connected to the Atlantic Ocean, its flow is subject to tides, and the water here remains slightly brackish. Low tide was at 11:19 a.m.; we had come an hour earlier, so that we would be able to stand up where the water deepens. The water was waist high or, at the most, chest high. The water chestnut stem can be several feet long. If you reach into the water and grab a plant by its stem, then wind it around your wrist before you pull it out, Nancy told me, you'll have a better hold on it and a better chance of getting it by the root. We filled our buckets this way, then walked through the water over to the shore, where we dumped them above the tide line where they would decompose. And then we did it a second time.

Nancy speaks with a quiet authority. She has been doing this for years. "There are some short-term things I do," she explains, "for two or three years, maybe. But this is something I come back to every summer. This is something that has to be done repetitively. There is a peace in this experience that I can keep coming back to." Nancy became a widow on September 11, 2001. Her husband, Jupiter Yambem, had been a banquet manager for Windows on the World in Tower One, and not long before that bright September morning, he had switched his schedule to a morning shift so that he could spend more time in the afternoons and evenings with Nancy and their five-year-old son. The couple had often crewed on the river and regularly assisted in weed pulls.

In the years since, she has remarried, changed jobs, moved, raised a teenage son whose passions range from skateboarding and guitar to civic activism. Her volunteer work has included serving lunch at the Salvation Army, teaching a Sunday school class, and running a bereavement group. But pulling up the weeds is something she returns to, summer after summer. Sometimes, as on this morning, it's a quiet enterprise planned weeks in advance. Other times, if it is especially hot and the salt front, the leading edge of seawater, has moved up the river early, there is a sudden call for volunteers to come down quickly and pull out the dying plants. The circumstances can vary. "Life does that to you sometimes," she says of that arbitrary nature of these things. "And you just know you'll end up doing it again. Just being on the river and sharing it, or thinking maybe I helped someone get to the river, there's a reward there, and the peace that comes with that. Sometimes it's a temporary thing, but it returns, in varying degrees, again and again. Just like the weeds."

Weeding a river is an exercise in which leisure and industry easily coincide; it's a brand of gardening in which a sense of purpose can intersect with being languid. From time to time, I saw an elver, a juvenile American eel, winding around a stem or root like some weird extra plant appendage. Although the fish diversity is lower here than elsewhere on the river, eels can withstand the low oxygen levels of the water chestnut bed, all the while snacking on its assorted invertebrates. Yet if the eels swim off quickly, everything else seems to take its own time. Like anything else that is done in water, weeding is done slowly, as though it is possible to take on the liquid motion of what is around you. The stems can be pulled out with the gentlest tug; their attachment to the riverbed seems slight, their resistance imperceptible. Yet there is the smallest bit of spring to them, as though some bit of elastic thread has woven its way through the watery pink tendrils, and they have that sense of give that the most tenacious opponents sometimes seem to have. With a bit of stretch, these interlopers seem to be hanging on, though without much faith. And the mud on the bed of the river has a give, too; at each step, we sink in a bit. Perhaps this is why I am so drawn to the waterworld of rivers: nothing here stays the same for too long; things are always shifting, drifting, gently giving way.

Our weeding accessories were primitive. Once we had pulled the water chestnuts up, we put them in small, gray buckets. Dime-sized circles had been punched into the buckets so that the water would drain out. The buckets were kept afloat with a bright blue foam swimming noodle encircling them and attached with a strip of silver duct tape. Buckets, noodles, tape. There was an almost child-like quality to the tools we were using, as though they were the accessories in some kind of water game. We were not dousing the leaves with pesticides, because there are no such things as safe pesticides. Nor were we relying on the blades of a mechanical weed cutter or harvester. They are expensive, and the heavy boats that carry them are impractical for shallow water. And while biological controls may be found in the future, there is nothing conclusive or licensed, for that matter, yet.

Not long before I had read an account of Qingdao, a city in China whose coast was being choked by an unexpected bloom of green algae, officially said to be a result of warmer weather and heavy rains, but more likely a product of excessive nutrients from sewage, industrial pollutants, agricultural waste. The photographs documented not so much a landscape of displacement but a grotesque parody of aquatic flora; what should have been the surface of clear water was a swirling, shaggy stew of chartreuse plant life. The city was to host the regatta for the summer Olympics, and thousands of city residents had volunteered—or been ordered—to go out into the Yellow Sea on foot or in small wooden boats to scoop up the green algae by hand.

And I wonder how it has come to be that our responses to these invasive species are so rudimentary—our little gray buckets, their little wooden boats; and how our ingenuity so often fails us when it comes to cleaning up after ourselves, as we are going out now, collecting these weeds with nothing but our hands. Certainly the term "citizen activist" describes the kind of effort we were engaged in this morning, but the phrase that stuck with me was "incidental steward": what we do and what we use to do it with both have the air of the improvisation that so often comes from necessity. Pete Seeger says the world changes one teaspoon at a time, but on the river on this shining July morning, it is more like one leaf, one stem, one seed at a time.

Despite the turbidity, it was low tide now, and easy to see the river bottom. The spit of land we were near was once the town dump, and I had seen a photograph in which plumes of smoke rose from what is now a grassy lawn. But the riverbed still bears evidence of that time, shards of glass, the remnants of cans and tools, bits of metal, parts of old plastic toys, and I was grateful to be wearing water shoes. Nancy found a dartboard, debris that looks more recent, but it was unlikely to serve again as any measure of human accuracy. I had always thought of a river as a place of running water, of currents and flow and velocity, a fluid artery that takes you from one place to another with a sense of purpose and direction. That afternoon, though, this quiet backstretch seemed more like an archive, some wide, watery cabinet that contained things: leaves, stems, pods, mud, floating logs, debris, rocks, tiny crustaceans, insects, mollusks. Things resided here. Amazingly, they could rest here. David Strayer, a freshwater ecologist at the Cary Institute of Ecosystem Studies in Millbrook, New York, has spent much of his professional life looking at the Hudson River, and he speaks of the way the river is familiar and mysterious at once. "Everyone knows what it looks like," he says. "They know where it is. And yet we don't know what's under the surface, what's living at the bottom, how substances move through the river, how all the parts work together to form the ecosystem."[2] Now, for a few minutes in an area of a few square feet, I felt as though I had gotten a brief glimpse into some part of that mystery.

And as we continued to pull the weeds out for the next hour or so, we found ourselves gently disrupting the arrangement of the riverbed, reassembling its articles in slow motion to compose a subaquatic still life that was only slightly different from the one before. A shoe was realigned next to an old bottle. A devil's head was pulled out of the river's silt. The situation of a shining leaf was changed. Pulling the weeds out of a river, I realized, is as close as you can get to weeding your dreams.

I wondered whether it would make a difference if only some percentage of these weeds was removed; and whether this is the kind of distress that can be alleviated by degrees. There are times, after all, when partial recovery is the most you can ask for. Can distress of this magnitude be thinned, pruned, diminished? Or, as a practical matter, must it be extinguished altogether? But in regard to water chestnuts, oxygen depletion, and maybe other things as well, it is the kind of idle speculation that is easy to engage in when you are spending the morning in and on the river. The only thing I was beginning to grasp for sure was what had brought me here, which had something to do with the chance to restore clarity to what is cloudy and cluttered in our lives. That it is possible to weed something so fluid as river water speaks to our ability to put some kind of order to those things in our lives thought to be too quick, too changeable, too transparent to require our care. Or as Nancy said, "It leaves you with a completed feeling, even though the job is never completed."

And so we pulled at the stems, and as we did, the shape was reconfigured, the space between the rosettes changed. I was reminded of the work of the artist John McQueen, whose materials tend to be those found in nature, willows branches, twigs, and weeds. But from time to time he also manages to find letters of the alphabet in these and in the limbs and leaves of trees, often photographing an A or an H or an M, a whole system of improvisational typography constructed by the arc and bend of branches and boughs. His search for these letters reminds me of the way we all look to the natural world for meaning. In the same way children find in a cloud formation a fish, a castle, sometimes even an entire continent, something they know, or perhaps, something they want to know, it seems to be in us all to look to nature to track something we might recognize, some elusive calligraphy of order.

But just as we can find meaning, so, too, can we remain indifferent to it. For all its elusive beauty, McQueen's woodland alphabet can't help but evoke the 1963 movie *It's a Mad, Mad, Mad, Mad World,* in which a trove of cash is hidden under the Big W, a grand configuration made by the sweeping curves in the trunks of palm trees. Something in me remains certain that the Big W is a precursor to the vast acreage of environmental art that has since been produced. Surely its message is as resonant as any piece of earth art by Robert Smithson or Michael Heizer. In the movie, hilarity ensues as Buddy Hackett, Milton Berle, and an assortment of other deranged characters keep looking around the palm trees, through them, in front of them, behind them, without ever actually seeing the extravagant letter they form swaying over their heads. I saw the movie when I was ten, and though I may not have realized it at the time, it taught me how comedy can sometimes be a product of nothing more than our scathing indifference to what is right before our eyes. And how laughter can sometimes make us weep. And as Nancy and I continued to reshape the contour of these leaves, revising the space among them, I found myself amazed all over again at how we seem to nurture equally and simultaneously the ability to locate a logic in the natural world and the aptitude to remain oblivious to it.

I have never been much of a gardener. I have nothing but respect for the process and certainly am grateful for the rewards—the Mexican pitcher full of daffodils in April or the bowl of fresh blueberries in August. But whether it is the time or the patience or the sheer volume of knowledge that seems to be required to cultivate a garden, I knew it was something beyond me. Each spring, though, it gives me pleasure and satisfaction to pull the weeds out of the perennial bed. There is something about yanking out the sprouts of maple, dandelion roots, and ryegrass that have threaded their way through the peonies and irises that answers to the part of me that wants to set things right. Pull out a tuft of crabgrass and get the better of an interloper. Tug the brambles out of the ground, and it restores some small sense of order.

Weeding this backstretch of river was a similar enterprise, but on a scale that is at once more grand and more elusive. And it occurred to me then that it is the opposite of an introduction, and I found myself wondering if there was a word for that. And where I might find the letters, possibly even an entire language, whether in the branches of trees or between leaves on the water or anywhere else at all, that might offer the word for that kind of polite disengagement that would enable all of us and the things we live with to become strangers to one another, to resume never having met. Surely that is what the situation required. I knew that if I ever found that word, though, it would disappear as quickly as it came, as fugitive as any cloud or leaf. All I had at hand, anyway, were the blue noodle, the gray bucket, the silver tape. And by the time the morning was over and the tide coming in again, I knew that weeding this river was probably as close as I'd ever come to knowing what that word is.

4

Pools in the Spring

Vernal pools come and go. They are one of those features of the natural landscape that are defined not only by physical characteristics but also by time, usually between a couple of weeks and several months. Seasonal wetlands covered by shallow, non-running water for variable periods from winter to spring, they're small, usually less than two acres, and not connected to other bodies of water. Fed instead by melting snow and spring rains, they often escape human notice entirely. Which may be why Ann, Joyce, Ray, and I found ourselves so baffled on that April afternoon. We were standing in a meadow at Fraleigh Hill Rose Farm in Red Hook, New York, looking and listening for some evidence of the vernal pool that, though indicated by the aerial map, was nowhere in sight.

If the identity of a vernal pool can be hard to fix upon, its effect on the landscape is more decisive. Because these pools are not fed by running water, they are not inhabited by fish, an absence that allows the pools to serve as secure amphibian breeding grounds. Relatively common and widespread species such as spotted salamanders and wood frogs breed here, as do the more rare and threatened Jefferson and marbled salamanders. All are players in a functioning ecological network, and all are declining due to loss of wetlands. Amphibians consume vast numbers of insects, and how they control the insect population has a direct effect on human health: without the pool, we are without the breeding ground; without the breeding ground, we are without the amphibian; and without the amphibian, we are without insect control, and another little piece of the natural balance of things has been lost. If there is any substantial distance between larger, permanent bodies of water, the vernal pools lying between them also provide resting and feeding grounds for a variety of other small mammals, birds, amphibians, and reptiles. Vital to the food chain, these smaller species may be unable to travel the longer distance without falling prey to larger forest animals.

Vernal pools are not simply habitats. They also serve that more basic function served by other bodies of water: they cleanse. The bacteria in their soil help to convert nitrates in the water, a result of common lawn fertilizers, into less harmful nitrogen gas that is released into the atmosphere. This denitrifying bacteria allow the pools to serve as small recharging stations for local aquifers, filtering and renewing snowmelt, rainwater, and floodwater of pollutants before they become groundwater. Vernal pools are efficient little natural water treatment centers, and were they to be compromised or to vanish entirely, the results could include flooding, a water supply both diminished and polluted, the loss of wildlife habitat, and erosion. The existence, size, and biological life of the pools determine their function in the landscape—some are robust habitats, others degraded and less useful. Assessing the presence, size, and biological life of these pools leads to reflections about how such shallow, seasonal, elusive features of the landscape can still have such a critical effect on it.

Local municipalities, though, rarely have the resources to hire environmental consultants to conduct such assessments. Although federal regulations have traditionally tended to govern decisions about larger wetlands, the existence and status of vernal pools often remain unknown to local and regional planning departments. In New York State, a wetland is subject to regulation only if it is larger than 12.4 acres, contains a state-listed endangered or threatened species, or is located within Adirondack State Park. Small, ephemeral features of the landscape, vernal pools remain largely unprotected, their management as elusive and unguarded as they are themselves.

It was in an effort to address this that Michael W. Klemens (in partnership with the Metropolitan Conservation Alliance, which he founded in 1997; the Cornell Cooperative Extension of Dutchess

County; and the Cary Institute of Ecosystem Studies) initiated a program to monitor the vernal pools in Dutchess County. A herpetologist as well as research and policy conservationist, Klemens approaches his work with the conviction that teaming up with local communities to disseminate scientific research and knowledge is a pathway to informed land-use planning. The first time I reached him by telephone, he described himself as being "interested in where-the-rubber-meets-the-road advocacy." He has been working with volunteer groups in a broad range of research projects since the 1980s, telling me, "This is an ethical calling. I am interested in what motivates people. So am I an ethicist, a scientist, an advocate? I went from being a research scientist to having a broader perspective. What do I do as a human being? How do I translate my knowledge to make a better world? How do you alter the trajectory? It has been a voyage for me."

Klemens's commitment to bring science into local land-use decisions isn't simply about training volunteers. It's based as well on the recognition that local residents have a sense of allegiance to the place they live, along with existing relationships with neighbors and landowners that can facilitate research immeasurably; such relationships are often what make it possible for researchers to conduct their work on private property. "Well-placed citizens can create the change we need," he says months later when we meet. "They can create small ripples in community."

Klemens and his staff had designated the town of Red Hook for vernal pool assessment for several reasons: the town board had expressed support for the survey and was likely to consider the information gathered in its future planning decisions. Red Hook is also home to Bard College, where environmental activism is embedded in the ethos and thus offered a significant volunteer base. Ann, the volunteer coordinator for our group, was a development manager and arts education coordinator at Bard College. With us as well was Joyce Tomaselli, a manager in business development and marketing at IBM whose job had been a casualty of the 2009 recession, and Ray Mansell, her neighbor. Red Hook had been divided into quadrants, and we'd been assigned three sites in the southeast quadrant. But the first one at Fraleigh Hill Rose Farm was proving elusive. Volunteers who had come to this site a week earlier had been unable to find the pool, and we weren't doing much better. A small section of red crosshatching on a printout of an aerial map indicated the presumed location of the pool, and we headed south along the edge of a young apple orchard to look for it. All we found, though, was meadow grass, a large pile of brush, a stand of pines.

Nothing, though, seemed quite as it should be. While spring in the Hudson Valley is a generally unpredictable affair, in 2010, the thermometer had hit the low nineties earlier in the month. Then it plummeted again, and most mornings for the rest of the month the temperature had fallen below freezing, though by midafternoon it might feel like August for an hour or so. It had been rainy and cool earlier in the day, but by the time the four of us met in the parking lot of the Cornucopia Discount Beverage Barn, it was in the seventies and felt like a summer afternoon. Now the freak hot spell during the first week of April had caused the apple and peach trees to flower prematurely, and their leaves were sprouting two weeks earlier than usual.

Ray spotted a fox skirting along the brush pile. We listened for the wood frogs, but couldn't hear those either. Joyce asked Ray whether he'd come across any morels yet. The poison ivy is just out, she told us, and the two usually arrive at the same time. One of those people who are able to identify what is happening in the natural world by looking for what else is happening, she is versed in phenology. Like anyone with an interest in natural history, she knows that when one thing happens, so does something else, and spring is a time when these simultaneities become especially apparent. We kept walking past the orchard and into a field with young blueberry bushes just coming into leaf. Moments later, there was a sudden flapping of wings. "A Canada goose," Joyce said to no one in particular. "Not Canadian." She is precise about identification, a particularity that is appropriate to the task at hand.

In her midfifties, Joyce had had a thirty-year career at IBM, and she has an affability and easy manner that resonates with confidence and poise; she is able to speak candidly about both her successes at IBM as well as her anxieties about the future of her professional life in a recession. Now she had thrown herself into volunteering and had little trouble being enthusiastic about making the transition from her work in global markets to volunteering for local agencies. "It's an interesting time now for people who have lost their jobs," she told me later when we meet for coffee. "There are the economics of need and a talent pool that can help."

Joyce used her old business card that lists her areas of corporate expertise: Solutions Business Development; Marketing Strategy, Plans and Implementation; Content Creation and Delivery; Launch, Go-to-Market and Channel Enablement; Enthusiastic, Knowledge Leader and Collaborator. It is a kind of language about aptitude and working skills that seems out of place in the woods, yet it's clear that she's bringing the same energy and exactitude to the pools and woodlands. Joyce may personify the emerging profile of the American volunteer, someone who has the desire and resilience to transfer the skills and knowledge developed in one area of life to another. As she put it, "I like to learn and know I have learned well. I try to be precise, thorough, to bring data that is valid, well organized, well documented, well labeled. I meet deadlines. I feel obligated to do volunteer work with the same core values." And if she was clear about bringing these same principles to this endeavor, her thumbnail portrait of her partners here reflected a like clarity. "The scientists may not be collaborative—they're not nurturers or gatherers," she said matter-of-factly. "But they have the precision to define the project well, along with the methodology to keep the project scoped and the volunteers focused."

We continued across the field of crabgrass, dandelions, a smattering of violets, looking for water and finding only meadow and fruit trees until it occurred to us all that we were participating in that time-honored human enterprise of searching for something that was not there. It was possible that the pool had been misidentified on the survey map. Ordinarily, the maps used to find the vernal pools are based on data gathered from remote-sensed geographic information systems that is then field-verified, but this year there hadn't been the time to check the sites. And even though the aerial photographs we were working with had been taken when the trees were bare and water levels high, their images could be obscure or deceptive. Cameras can deceive, and here, shadows and gaps in the tree canopy from timber cutting can suggest that a pool exists where, in fact, it does not; likewise, an existing pool might be missed because of its small size or canopy cover.

In studies using satellite imagery and aerial photographs, the term "ground truth" refers to what is actually there rather than what appears to be there on the pixels or whatever other data information system is being used. Ground truth means what is "on location," what is real, what your eyes and ears tell you rather than what the sensors suggest. In any kind of data collection that uses remote sensing, it is vital to compare remotely gathered information with what is found on site.

And I wondered now whether this quest for ground truth, or something very much like it, is something in which we are all increasingly engaged. Maybe it implies the same difference between pixels and brushstrokes, between the aspirations of a Facebook profile and what might be revealed over a bowl of soup at dinner, two different kinds of information, one abstract, selective, filtered, the other rooted more solidly in physical truths, firsthand observation, and direct experience. At the best of times, these two realms of information support each other; sometimes, though, they negate and defy each other altogether. Whether it is the pixels of electronic imagery or the fragments of information on a social networking site, the digital world gives us its truths. Yet how we adjust these to what we see with our own eyes is a process that demands its own thought and imagination. It requires a kind of thinking, a kind of intelligence, possibly a kind of ingenious reconciliation we have not been called on to practice before. Now, on this spring afternoon, the four of us confronted the ground truth that what we were after might not exist at all.

A five-minute drive took us to our second site. Number 249 was much closer to the road, and volunteers who had come to this pool a week earlier had found four egg masses for spotted salamander, along with two redback salamanders in the water. Frogs jumped into the water at our approach. A lacy scrim of duckweed covered the surface of the water. Ray exclaimed at the tadpoles, invisible to me until he handed me his polarized sunglasses, and then I could make out the hundreds, maybe thousands of tiny black squiggles in the water.

And a few steps beyond we spotted a large egg mass floating near the shallow edge of the pool. It was a watery globe, possibly three inches in diameter, in which pale green eggs were suspended. Joyce identified it as spotted salamander. Ray gently lifted the egg mass out of the water and held it while she photographed it. Then he handed it to me, and for a moment I found myself the steward of something utterly unknowable. How often we feel as though we are custodians of uncertainty, how often we find ourselves holding the unknown in our hands, yet on this spring afternoon, as my fingers formed a basket for this amphibian egg mass, it was really, manifestly true. For sure, it had been identified: we had been instructed about its color, shape, size, the number of eggs per mass; we had the color photographs to compare it to for identification; we had its written description; and we knew that we had found it where it was likely to be found. And yet. There was something in this amorphous cloud of jelly filled with eggs that rendered it the material expression of uncertainty. Quickly, I placed it back into the pool.

And then we looked for more, stepping around the skunk cabbage and winter's fallen boughs and limbs. Joyce identified trout lilies, false Solomon's seal, then shoots of snake grass, a slender weed with an articulated stem, and early blooms of coltsfoot, ticking off their names in a way that suggests that the arrival of spring comes with a checklist; her impulse to name things may be a way to claim them. Perhaps more than that, correctly identifying something may be the first step in knowing its larger place in life. For all the precision of her list, though, what was submerged in the water was more obscure, camouflaged by sodden leaves, decaying branches, and the dark water itself. As we skirted the pool, we found two more egg masses, then a fourth, and I realized that finding the first has recalibrated our attention; we have fine-tuned our watchfulness. As in so many human enterprises, find one bit of what you are looking for, and an extravagance of possibilities suddenly opens up.

Searching beneath the surface of the dark water for egg masses requires a different kind of attention. I know that generally I am not a reliable witness. What appears in front of my eyes can vanish in an instant. I had learned this several months earlier when our house had been broken into. I had surprised the intruders, even spoken with them briefly, then watched as they fled quickly in their car. No one was hurt, nothing was taken. Yet when the police questioned me later, my recall was scant. "It was a small, gray car," I told them, next to useless information. I had not registered whether the car had headed north or south out of our driveway. It was a New York State license plate, but not if my life depended on it. The man could have been twenty-eight or thirty-eight. The woman was wearing sweats, no jeans . . . I don't remember. "She was short." "How short?" "I don't know." I was a witness without information. And they had been on the verge of robbing me.

Maybe now, four months later, my standing at the edge of this dark pool on an April afternoon was an effort to recalibrate my attentiveness in some small way. To discern what has been taken and what has been left in a different manner of theft. To identify who has had a hand in it. I want to know what is happening around me.

And yet I am familiar with the quandary inherent in attentiveness. Close observation requires that we choose what to observe and what not to observe; attentiveness to one thing so often mandates inattentiveness to another. Maybe it is this paradox of watchfulness that I am hoping these outings will make clear to me.

Oblivious to property lines, the pool straddled two lots, and we had received permission to look for the egg masses from only one landowner. After we had been there for a few minutes, she drove up in her Subaru. "What are you going to do with the information?" she asked. That is always the question. There are probably few of us who would admit to wanting to diminish the ecological health of our land; it is innately human to want clean water, trees that thrive, earth that hasn't been contaminated by toxins. Yet instilled in us as well is a desire to determine what we build and where and how, and a distrust toward those regulations that might restrict our desires.

It is exactly this paradox that Klemens's work with local planning groups addresses. "Ownership is about rights and responsibilities," he said to me months later when I met with him at his home in Connecticut. "We are good at defining the rights, poor at defining the responsibilities. And land is the last frontier where public interest is not well defined, unlike art, say, or historic buildings, where public rights often trump individual rights. Property rights bring stewardship responsibilities. And we have to learn to use property for the betterment of the common good." Klemens's geothermal passive solar house is built just above the floodplain where the Housatonic and Hollenbeck Rivers meet, and before we talk, he has taken me on a tour of the area. To stabilize the banks of the river he has been planting trees, bur oaks, willows, red and silver maples interspersed among the native box elders and lindens. The meadow that had once been a cornfield leeching nutrients into the river is ablaze today with goldenrod, clover, bergamot, asters, Queen Anne's lace. "Here I have zoning restrictions and conservation easements," he says. "I can't do just anything I might want to do. Ownership is not only about the here and now. It's also about the past and about the future."

The woman in the Subaru repeated her question, but it was one that had not yet found an answer. It was 5:30 p.m. by now, and a half moon had appeared overhead, emerging from the tulip trees and ash, maple, white oak, and pin oak at the edges of the pool. We listened to the steady chirping of wood frogs. No adult salamanders.

The third visit took us to two interconnected pools, Numbers 310 and 339 at 243 Stone Church Road. As soon as we got out of the car, Ray spotted a Cooper's hawk, and I realized that, throughout the afternoon, he had been constantly murmuring about the birds, the recitation of what he heard a soundtrack for our outing. He understands the world through sound, taking it in through its acoustical signals; he listens.

This was the third visit to these two pools. The volunteers who had made previous visits earlier in the month had found nothing, and today, the thin veil of duckweed over the surface of the water in the first pool made it nearly impossible to see anything. It was one of those moments when the essential information seems to be available, though just out of reach, something like looking at a page of indecipherable script or the clouded face of someone who is unable to explain himself or herself. The natural world also manages to withhold information, practicing its own reticence. Though we could hear a wood frog, we saw nothing. It seemed hard to believe that such a dank, opaque pool could function as an agent for cleansing and renewal. Vernal pools cleanse groundwater. Water cleans water. The dark pool offered nothing more than what seemed like an impenetrable Zen parable.

The connected pool some twenty feet away was smaller, and we hiked through a thick stand of common reed to get there. Fast growing, dense, likely to crowd out other plants, and diminishing species diversity, the reeds are rarely welcomed in wetlands. Here, in their persistent stranglehold across this small marsh, the dried and withered stalks, along with the new young, green shoots, camouflaged two old garden chairs left by the pools, fastening them to the ground. The chairs, one white plastic, the other wrought iron, seemed some portrait of antique vigilance, a monument to some kind of benign watching, before global positioning systems, aerial survey maps, and data sheets. "A Carolina wren," I heard Ray murmuring to himself.

Joyce found a deer path, and though it was choked with wild blackberries, it took us to the second pool. It was early evening by then, and a brown bat surprised us by flying out of the brush. Filling in the data sheet for the first pool, Ann said softly to herself that it was a dead loss. Then added to no one in particular, "Except for the bat, the wood frog, the Carolina wren. So, really, it's not a dead loss."

And then right away when we got to the second pool, we found two egg masses, one lying on the bed, the other attached to a branch submerged in the water, both surprises because the two previous visits to this pool had turned up nothing. Milky, white, kidney-shaped, these were the egg masses of the spotted salamander, the blob architecture of the natural world that can attach to sticks like bulbous flags, luminous, vague, imprecise, and indistinct.

At the far edge of the pool were the remnants of an old stone wall, now barely visible, and elsewhere the thick underbrush had made the edge of the pool inaccessible. Poison ivy was a further deterrent. There was no way to search out whatever other egg masses may have been there. I thought of all the ways there are to come up with nothing. You can't see it or you can't get to it, your boots aren't high enough or the water is too deep or dark, the wet leaves are hiding what's there, the branch across the water obscures the bottom of the pool, the duckweed is too thick—there are a thousand ways to come up empty, ten thousand ways not to see what is right in front of you. But then Ray heard a wood frog and saw another one tucked under the roots of a tree at the edge of the pool. Frogs are small totems of good luck in Japan, especially for travel across water, and although there was no place on the data sheet for this information, I suddenly felt sure that such enduring associations with the amphibian world had helped to bring us to this pond.

Volunteers provide an economical human resource, and there is strong research supporting the idea that data on amphibian egg mass counts provided by adequately trained citizens does not vary significantly from that gathered by biologists.[1] Still, there is a limit to what we can do during our afternoon in the woods. Along with the number of egg masses and the condition of the pool, scientists need to know the pool's effect on its immediate landscape; a 100-foot radius around the pool provides nutrients to the pool's ecosystem, while the area between 100 and 750 feet around the pool is a critical upland habitat used for foraging, shelter, and hibernation. Evaluating that extended habitat asks for a level of skill and training beyond the scope of volunteers and is better left to trained ecologists. But it is exactly this coalition of effort that Klemens supports.

His conviction that "the model of the detached, priestly scientist" is outdated and irrelevant sometimes puts him at odds with the research community, but his engagement with local community groups and planning agencies brings different recompense. "Advocacy for scientists is not in the training and it is not rewarded," he explains. "I am not asking for an epiphany in the scientific community. I'm just asking that they don't stand in the way."

Klemens questions the time-honored boundaries between science and advocacy, suggesting that in urgent times such as these, the need for information is no less critical. The scientific community has traditionally viewed its work as the research on which application and policy can subsequently rest; scientists are often reluctant to campaign for public policy themselves, because ideological sponsorship of any sort can put the integrity of the research into question; a strong bias colors the science, or so the argument goes. It is Klemens's hope to move beyond this, toward an approach that is often called "post-normal science." The term was coined by authors Silvio O. Funtowicz and Jerome Ravetz, who state that "the science appropriate to this new condition will be based on assumptions of unpredictability, incomplete control, and a plurality of legitimate perspectives."[2] Or as Klemens says, "It's not a question of being sloppy in science, but rather, the idea that robust data can be sufficient to inform policy. The levels of certainty do not have to be at 95 percent. It can be sufficient to have a robust data set regarding cause and effect or probable outcome to inform prudent public policy."

An afternoon at the pools is a good lesson about the enduring effects of the ephemeral and how something so fugitive can hold so much critical information. Why does it matter if some of these species are lost? Human health depends on species biodiversity. There is a direct correlation between diminished biodiversity and increases in infectious diseases. Or in the words of Edward O. Wilson, the loss of species is also a loss of information.

> Vast potential biological wealth will be destroyed. Still undeveloped medicines, crops, pharmaceuticals, timbers, fibers, pulps, soil-restoring vegetation, petroleum substitutes, and other products and amenities will never come to light. It is fashionable in some quarters to wave aside the small and obscure, the bugs, and weed, forgetting that an obscure moth from Latin America saved Australia's pastureland from overgrowth by cactus, that the rosy periwinkle provided the cure for Hodgkin's Disease. . . . In amnesiac revery, it is also easy to overlook the services that ecosystems provide humanity. They enrich the soil and create the very air we breathe.[3]

That night the faint red crosshatching that indicates the vernal pools on the aerial survey maps appeared in my dreams. The pale grid did not materialize as a graphic device but had morphed instead into a web of sorts, a fragile safety net that had been stretched across the subtle depressions in the ground, and in my dream the search for these nets became a quest for some kind of greater security.

As it turned out, though, the information held in the pools was almost as elusive as the pools themselves. Lack of funding had made it difficult to fully develop maps of the pools before assessments, nor could those that *had* been mapped be field-checked. The first pool for which we had searched with such diligence along the edge of the apple orchard and by the stand of pines was probably nothing more than a digital apparition. Neil Curri, the senior geographic information systems resource educator on the project, later told me that along with aerial photos, he had tried to use material from the national wetlands inventory to identify the possible location of pools for the survey. But even then, it is an unsure exercise. "Areas that have semipermanent hydro periods may be indicated but are not necessarily vernal pools," he said.

Organizing the roles of volunteers and coordinators had been difficult as well, and when data sheets were finally delivered, further budget cutbacks had prevented immediate assessment or analysis. When I asked Curri about seeing the results of the assessment, he sounded fatigued by the process. But I wanted to tell him that that searching out a vernal pool on an April afternoon is, by nature, an exercise in ambiguity. The entire enterprise is one of uncertainty—which has its own value. Just as a pool that is temporary, obscure, constantly shifting in its size, shape, and depth can decisively determine what occurs on the ecosystem around it, so, too, can the less quantifiable experience of those spring afternoons have lasting meaning. It's not necessarily about what you find; sometimes it's about being present, showing up. Or as Klemens says himself, "When people are close to the land, they can make decisions that are informed. They understand the consequences. In a technological society, we can get so divorced from nature and don't understand the consequences. When people experience it firsthand, they can advocate for it."

It was the end of the afternoon by then, and we walked back to the car, stepping through Jack in the pulpit, the tiny baby spring ferns that none of us could give a name to, the scattered wild geraniums. A low warbling filled the air. "It's a family of bluebirds," Ray said softly.

As we were getting ready to leave, a man pulled into the driveway in his truck. When he got out, we paused just long enough to tell him what we'd been up to, about the egg masses in his pools, the two wood frogs, one seen, one heard, and the sudden swerve of the brown bat.

He nods and looks pleased. "That's great," he says, "but I'm just a tenant."

5
Ribbons
Underwater

Mid-August in the Hudson Valley is a good time to take measure. The fields are a riot of wild carrot, chicory, goldenrod, and joe-pye weed, and the woods have become an impassable thicket. It is no different on the Hudson River. Until a dry spell allows the salt front to move upriver and make its assault on the water chestnuts, the watery meadows of rosettes flank the shore, and at Stockport, the mud flats are a pasture of spatterdock at low tide.

The plant life is just as abundant beneath the surface of the river where the beds of wild celery, *Vallisneria americana,* flourish in late summer. Unlike the water chestnut's rosettes that blanket the surface of the river, wild celery remains submerged. Its leaves, slender strands, permit light to penetrate the water, even as they lie flat at water's surface during low tide. Wild celery oxygenates the river, offering food and habitat to such invertebrates as aquatic earthworms, scuds, insects, clams, and mussels and to fish such as perch, carp, and eels. Canvasback ducks dive for food in the beds, and egrets and herons forage there as well. All of which is why a dense bed of it speaks to the health of the river. The beds occupy about 6 percent of the river from Haverstraw Bay to Troy, but where the plant proliferates varies over time. *Vallisneria* is a tough perennial with belowground reserves. A tuberous structure under the sediment, part of its root system, gives it a certain persistence. That said, salinity, pollution from combined sewage outfall areas, and marine traffic can all work to diminish it, while sharp shards of winter ice can dig up the sediment in which it grows.

Taking measure of wild celery, then, is a way of taking measure of the river's health, and protecting its beds from human activity a way of safeguarding river ecology. But calculating this submerged waterworld presents practical problems. Every five years since 1997, aerial photos of the Hudson have been taken to document the river's submerged aquatic vegetation, but such photography is costly. And although these photos can capture the outline of a bed with tone, color, shape, size, and texture, glare sometimes obscures the image. Photographs are also unable to record the depth of the water or its

Ribbons Underwater

clarity, other factors in how *Vallisneria* grows. Add to that the constantly changing condition of the beds, which vary in their growth and density, and it is difficult to know what is going on beneath the surface of the river from year to year. To find out, a broad partnership of scientists, educators, and resource managers from the Cary Institute of Ecosystem Studies, Cornell Institute for Resource Information Sciences, New York Sea Grant Extension, and Hudson River National Estuarine Research Reserve established a volunteer monitoring program in 2003.

Now, every summer between mid-July and mid-September, Stuart Findlay, an aquatic biologist at the Cary Institute, works with a geographic information systems mapping specialist, an aerial photographer, and a volunteer coordinator, along with some twenty kayakers and canoeists who have been trained in data collection, to take stock of what's happening out on the river from the Tappan Zee Bridge north to the Troy Dam. It is not simply a question of taking inventory but one of tracking changes in the growth patterns of the plant. Equipped with maps marking out transects and location coordinates, along with handheld geographic positioning systems units that help to locate these coordinates, volunteers go out in pairs to locate the beds and determine their density, along with water depth and water clarity. But the objective of the program isn't simply to gather data and compile statistics. Although local planners will refer to such information when making decisions about land use and waterfront development, the monitoring is also a manner of outreach that connects communities to the river that runs through them.

On this August afternoon, I was going out with my friend Doug Reed, an environmental educator who had been active in the monitoring program since it started. Doug is the director of Hudson Basin River Watch, a citizen monitoring network that he founded in 1993. The watch's mission is to improve the river's water quality through education and community involvement. In the years since then, he has worked with schools up and down the river, teaching teams of students to survey, sample, and analyze the water in tributaries from Westchester to the Adirondacks. In 1999, the organization developed a how-to manual to standardize citizen monitoring throughout the watershed, and these protocols have since enabled its residents to become partners in how river policy is made; they have identified out-of-date sewage treatment plants, reported polluting landfills, and otherwise become stewards of the water—and land—in their communities.

Doug gets around the river in a wooden rowing canoe, handcrafted with slender planks of hand-bent and laminated pine and finished with black cherry trim. "It feels like rowing a cello," he has said of its lightness as it passes across the water, and now, as we head out onto the Hudson, an angler on the shore notes its elegance: "Almost a shame to put it on the water," he calls out. "I'd just leave it on the coffee table and look at it." The canoe is outfitted with woven cane seats, and on a particularly hot day earlier in the summer, I had seen Doug rowing with a stylish black-and-red umbrella propped up over his head. With his sculptural handmade boat, fluid oar stroke, thatch of silver hair, and Panama hat, he brings to the entire enterprise a grace that seems exactly suited to natural beauty of the location itself. It is as though his boat and he have had the courtesy to dress for the occasion, an outing on one of the country's grandest rivers. Accompanying him on the Hudson in my little polyethylene kayak with its splashy turquoise graphics is something like being the hick cousin who, in all her K-Mart finery, inexplicably finds herself going to the ball with the count. But no matter. Doug's graciousness has a broad reach. "I love your kayak, big bad blue paint job and all, really!" he has assured me.

Low tide was just after 6:00 p.m., and because the wild celery is more visible in shallow water, our time frame was limited, only a couple of hours before and after that. When we set out a little before five, it was still in the mid-eighties and the water like glass. We pushed off onto the shining concourse just as a constellation of gulls skittered across the mud flats to the south; to the north the first great blue heron of the afternoon was poised motionless on a bed of water chestnuts. Doug had entered the coordinates for our waypoints into the global positioning system unit, and we paddled in the direction of our first transect, which meant finding the three or four or five assigned waypoints, then dropping in a small orange float at each to mark it. The invisible line these form make the transect where we would check for vegetation. Yet it was just as Doug observed: "The computer lays out the transects in a straight line, but it's never as straight on the river."

The manual for this project includes a twelve-page appendix on how to use the Garmin software, with instructions about downloading waypoints, putting them into text files, and then uploading these into the handheld units, all of which entails menu options, opening ports, auto connects, and projection zones. At a training workshop earlier in the spring, I had done my best with the Garmin to find arbitrary waypoints at assorted outdoor locations near the Cary Institute's parking lot, under the spruce tree, over by the edge of the road. The frustrations of that afternoon were nothing compared to those experienced in trying to identify a waypoint on a moving river. Water changes everything. Stuart had said as much at the training session: "75 percent of it is just getting there. Getting the data is 25 percent," and I saw now what he meant. With the pull of the tide and the current, the faintest surface breeze, the inevitable drift of the canoe, or any of the other elusive factors that figure into the way a boat passes over water, the global positioning system coordinates were fugitive at best. Whatever happens out on the water follows a less direct flow line.

Research in the computer industry is aiming to make these devices more sensitive, and in a year or two Doug may negotiate our transect along the river on a touch screen rather than by punching in numbers. Perhaps such a device that is more responsive to human gesture, touch, and intuition would make it easier to align our water path with the coordinates we have been assigned. A surface wind on the river was enough to remind me that how we know where we are and gauge our position is always an elusive exercise. Small wonder that the data sheets asked that we document both the expected and actual locations.

It's not just the coordinates that were hard to pin down. The river world is always ambiguous, and for the proliferation of all the water weeds this time of year, for the acres of spatterdock spread out across the flats and the wide beds of water chestnuts, the Stockport flats are a curious place to take an inventory. The river here has been completely reconfigured, engineered, and reengineered from dredging decades ago. Continually washed by four-foot tides, the islands, flats, and marshes are the result of dredge spoils where the shallow-water areas of the river were recontoured in the nineteenth and early twentieth centuries to accommodate large ships traveling up to Troy and Albany. Within only a few feet, the depth of the river here can change abruptly, from four inches to over twenty feet. Out in the channel it is deeper yet.

But where we were laying out transects, the river went from shallow to shallower and back again. The tide only added to the changeable boundaries of river and flats; one became the other quickly and constantly, and the contours of what was solid and what was fluid, what was water and what was ground, were continuously changing. Surely it was folly to imagine taking measure of such a place.

By the second transect, then, we were out of our boats and wading. The river was four inches deep here, too shallow to paddle, and Doug got out of his canoe and just pulled it along, wading between the waypoints. For all the clarity of the water and brilliance of the afternoon light, trudging through the shallows, along the meadow of mud flats and spatterdock was a clumsy exercise. Along with the orange buoys, we were equipped with a Secchi disc, a black-and-white circular form that is lowered into the water on a string marked off with centimeter markings. Determining how deep the disc can be lowered into the water before it is no longer visible is a way to measure clarity, which can be affected by sediment, precipitation, and wind stirring up the water. With no place for it to go here, the disc stayed in my kayak.

Several dark-shaded patches on the 2007 aerial photographs had suggested the existence of *Vallisneria* on this spot along the river; a vague triangle or blurry linear strip might indicate an area of growth. But the images were laid over one another, square over square, in an overlapping vertical arrangement that was oddly static and had little to do with the wide continuum of the river, seamless and ceaseless on a summer afternoon. Not for the first time I wondered about the wisdom of trying to match up the natural world with such obscure digital images. Doug laid out the third transect, but still, nothing. The data sheets offered a range of choices for the density of the beds—none, sparse, average, or dense—and I continued to write repeatedly *none, none, none.*

But none, I knew, is still data; zero can be data, too. I had heard this statement more than once, and it's a good lesson: when we don't see anything or hear anything or say anything or think anything, when we have nothing to go on, that, too, is information. This seems to be a known fact that transcends the boundaries of discipline. In music, silence can be as resonant as sound; in mathematics, zero can have the authority of a number. Robert Ryman's white canvases are a primer in the richness of white space, Merce Cunningham's choreography an inquiry into stillness. *Waiting for Godot,* one of the most celebrated plays in the English language, is famously described as the play in which nothing happens. "I have always been as interested in what is not there as what is there—the void, the interior spaces, the things that you don't see," the designer Marc Newson has said.[1] And in the words of abstract painter Ellsworth Kelly, "The negative is just as important as the positive."[2] Absence can be as powerful as presence.

Now, though, another reward offered itself. A bald eagle was perched on the upper limb of a tree on the flat just to our south. I tried to whisper to Doug, but he was too distant to hear me, and I stayed in that small hell of someone who has good news but can't share it. Doug is practiced at what he does here, which is to say he knows how to *observe.* He can take in the water depth and clarity, the density of the *Vallisneria* all the while remaining attentive in a more peripheral way to the larger picture; while marking out transects, precise in noting time and measurements, he is nonetheless able to take in the egrets, a great blue heron, a deer grazing on the shore with two fawns. He sees the details as well as the context, which is exactly why there was no sense or justice in his missing the eagle that sat motionless on the upper limb of a tree. Without warning, then, it was airborne and glided upriver.

We headed downriver, and though there was still no sign of the wild celery beds, Doug spotted a cormorant with a strand of it streaming from its beak. With its long, curved neck, the double-crested cormorant is graceful and goofy at once, and this one appeared to be floating downstream with a smug satisfaction at having come upon the *Vallisneria* before us. "That's what has happened to all the wild celery," Doug sighed. "The cormorants have eaten it. It's a new theory. Let's leave it to the scientists to figure it out."

Not until the fifth transect did we finally spot the watery tendrils, and then, suddenly they were everywhere. Where they brushed the surface of the water, they lie flat, combed by the current in an infinite variety of patterns, but beneath the surface, their dense growth was even more apparent. Now, as my kayak slipped across the water, it made a soft rustling sound.

We had found what we wanted where we wanted it and when, an alignment of desire, expectation, and discovery that I recognized to be an infrequent event. Wild celery puts its roots in the riverbed, but its leaves grow as slender ribbons, no more than a quarter-inch wide but up to seven feet long. If the water is dark and visibility low, it's easy to miss them altogether, to paddle across a bed and never know they are there. Looking closely now, though, I could make out their empire of strands, a vast submerged spreadsheet of the river's health. The water had a different texture here. There was something in the exercise of locating transects, about calling out the location numbers that seemed to be a way of assuring ourselves there is an order to what is happening out here, a way, perhaps, to catalog and identify the unseen. But the ribbons drifted and flowed in whatever way the current combed them, marking out a different, more tenuous transect. By now I knew what ground truths were and could identify their equivalents out here on the river. Water truths, by nature, seemed to be transitory and elusive.

In 1977, the artist Walter de Maria installed his work *The Lightning Field* in a remote desert area of New Mexico. The piece consists of four hundred rods of polished stainless steel embedded in the earth in a vast grid. Some twenty feet tall, the poles invite lightning during storms. But most visitors are said to find the arrangement of rods just as compelling on clear days; perhaps it is enough to witness the possibility of having these poles connect the wide desert sky with the red earth. Now I imagined that this subaquatic meadow was some remote corollary to de Maria's field and that these fluid tendrils existed in some parallel universe to his steel rods, some alternative connection not above ground but beneath water, vegetable rather than mineral, drifting randomly rather than on a grid.

For the next hour, we found ourselves navigating the shallows, at times noting abundant and dense beds where only a few feet away there was nothing. The categories offered by the data sheet are accompanied by drawings and corresponding percentages: none, 0–10 percent; sparse, 10–15 percent; average, 25–50 percent; and abundant, over 50 percent. I understood now that the word "data" is derived from the Latin word for "give." The weed can flourish in water that is up to six feet deep at low tide, and here, where the depth of the river seemed to change so quickly and unpredictably, determining its exact location was just as erratic an exercise. At dusk, just after low tide, the river was dark glass. By six-thirty that evening, the river was smoother still but the air suddenly cooler. The sun setting behind clouds gave the river a pearly sheen, and within minutes the trees on the western shore were reduced to a dark bank of foliage. The sky was as changeable as the shoreline. Gray rain clouds billowed to the north, but directly overhead was an iridescent white cloud.

There can be something mesmerizing about taking these measurements, and I knew then that it was not simply about taking measure but about *how* we take measure. I was reminded of all the different ways we have of counting—some abstract, theoretical, remote, others, like this, a direct, immediate, physical act. Our choices seem to have grown. I gauge books in pages, unless I am reading on a Kindle, in which case I am compelled to think in percentages. And then there are those calculations that have no limits. My friend Romig, a potter, has an order from Korea for ceramic tableware, hundreds upon hundreds of plates, cups, bowls that she tells me she will be unable to fulfill in her lifetime.

And my friend Myron has looked to a different lesson plan in the immeasurable. A gardener, he has recently taken to planting saffron crocuses. The small lavender flower blooms for several weeks in the fall, at which point he takes delight in using the word "harvest" to describe the process of pinching out the three spiky red stigmas produced by each flower with his fingers or, sometimes, a pair of tweezers. He dries the small threads on a piece of paper, then stores them in a dark place. When he offered to show me his crop, I saw six strands on a small sheet of paper; beneath it was another sheet of paper with three strands. Five thousand, four hundred flowers will be required to produce the 16,200 strands needed for a single ounce of saffron. How many Octobers would this take, I wondered. Jorge Luis Borges once noted that it was inconvenient to deal in inconceivable numbers, but convenience is relative, and all Myron said was that he would invite me for a bowl of paella when he had enough threads to flavor a pan of rice.

Since 2003, volunteers have collected twelve hundred data points. By the end of the day Doug and I will have added only twenty-five or thirty more. How many will be needed? I remember Nancy speaking of leaving the river after a morning weeding out water chestnuts "with a completed feeling even though the job is never completed." Maybe in the twenty-first century, the collective effort to compile a comprehensive data set is the equivalent to those labors made by previous generations to build pyramids and cathedrals, stone by stone, beam by beam. Possibly all of these fall into that category of enterprise alluded to by the theologian Reinhold Niebuhr when he observed, "Nothing that is worth doing can be achieved in our lifetime; therefore we must be saved by hope."[3]

But the thought, a bit grandiose for an afternoon out on the river, fled as quickly as it came. Besides, the way we were taking measure had a texture and light and current of its own. Maybe, too, it had to do with the simple repetition, the ritual of doing the same thing over and over again, though the results are always different. Or maybe it had to do with recalibrating a sense of time. In an age of acceleration, things seem to happen instantaneously and we react in like fashion. With our instant messages, thirty-second boot-up times, and drive-thru pharmacies, life happens fast. But it is possible that our transactions with the natural world can readjust our sense of timing. Such transactions follow a different cycle, whether minute to minute, year to year, decade to decade. Laying out the transects, determining the depth of the water, calculating the density of the wild celery—all of these seemed a way to reset the count, realign our sense of measure.

Which is why it was hard to stop, and even when a three-quarter moon slipped up over the trees on the eastern bank, we just kept going downriver, figuring we had time to just do another and then another, even though we had started to lose the light. That desire to follow that line down the river, to just keep going, to drift down the reach of the Hudson, taking one more depth measurement, testing one more time for water clarity, seemed to follow the course of some different transect. At 7:10 p.m., the river took on a golden sheen. The tide was coming in now, though we could barely feel it. At the tenth transect, we found point 4683641, but only a moment later Doug's canoe shifted off the coordinate. Watching him pull quickly on the cherry oars while his eyes were focused on the Garmin was like watching a choreography of tools from different universes. Maybe that's what it takes. By the time he took the depth measurement we were at 651. "Well, it was 641 when I started," he said matter-of-factly. This is how shifts occur on the river, quickly, quietly, without your ever noticing.

We had collected 29 data points. Months later, after Stuart has compiled and analyzed our data sheets, they will be added to the 346 data points gathered from fifteen other sites. And while Stuart had expected 38 percent sightings of *Vallisneria,* the actual figure is 43 percent. Data collected by volunteers in 2007 and 2008 had indicated a diminishment in growth, but the plant now seems to have rebounded and the inventory gained efficiency. The New York State Department of Environmental Conservation will use the information as it evaluates permits for waterfront construction, but beds of submerged aquatic vegetation have already figured into land-use policy when developers of a marina project in nearby Troy were denied a permit based on *Vallisneria*'s robust presence. The data also point to a connection between areas of the river with subaquatic vegetation loss and banks reinforced with hard engineering (using such materials as steel and concrete). The information will be of value as communities up and down the Hudson reevaluate how to modify shorelines as necessitated by climate change and rising waters.[4]

Tonight, though, identifying what was happening out here included a more varied set of calculations: along with the waypoints, the transects, the readings for depth and clarity, and the estimations of density, there was the arc of the snowy egret's sudden lift, the changing slant of afternoon light, the rate at which the trees were taking on shadow, and all the other factors of order on an August evening. A great blue heron standing on the flats captured the shade of evening and was bluer than ever, almost indigo, but its beak caught the last of the light as the sun slipped away. How little of the information that surrounds us is quantifiable. And I wondered then if what we had on our hands was an extravagance of coordinates. The thought vanished almost as quickly as it came, and we headed upstream.

The river was a dark sheen and caught only a bit of what golden luster was left in the sky. We paddled upriver effortlessly, the incoming tide gently reaffirming our direction. Ominous thunderheads had darkened the sky to the north, but summer storms tend to come in from the west, and foolishly I resisted looking at the sky upriver. Besides, we had been lucky so far. So when the surface of the water was suddenly furrowed and I could no longer deny the surface wind and ripple to the water we were paddling against, I knew it wasn't the tide but the storm coming downriver. The rumbling was getting closer and the darkness was no longer just the sun slipping away but the rain clouds bearing down on us.

We'd been hugging the flats on our trip upriver, so when the storm met us, we got out and pulled our boats onto the island. We took what cover we could below the tree where the eagle had rested, and suddenly it was raining hard. This summer shower would pass within minutes, but Doug hauled out the wet-weather gear and we waited it out, pelted by rain, on our little scrap of mud flat. When the lightning cracked the clouds apart, I thought again of de Maria's acreage of metal rods fastening themselves to those bright electric threads and how even on a clear day they hold out some gleaming possibility of a connection between earth and sky. But today the meadow of ribbons in the river offered a force field on some parallel plane that hinted instead at some connection to what lies beneath the surface.

The sky remained a dull steel gray to the east, and though we heard the rolling thunder, it wasn't far back to where the Stockport Creek emptied into the river, and we paddled back just before nightfall. After we'd strapped my kayak to my car and Doug's canoe to his truck, I handed the data sheets, not quite soaked, to him, and he put them in his truck. Months later, he had occasion to show them to me and pointed out that while I'd been accurate about water conditions and clarity, about the density of the *Vallisneria*, and precise about the hour and minute, I'd gotten the month all wrong. In some cases the shorthand on the data sheets suggested that we were on the river on a June afternoon, elsewhere October. Only a few of them correctly noted that it was August.

It's probably nothing more than the time travel that comes naturally on a summer afternoon on the river, Doug gamely suggested. While I knew he may be right, it occurred to me that for all those ways we have of counting out the things we see or do, such equations are easily forgotten when you want them to go on forever.

6

Coyotes Across the Clear-Cut

The small building I use as an office is up behind the house at the edge of the woods. It is at that line where the lawn meets the trees, where a patch of long grass, timothy, fescue, and strands of golden-rod separate the mown grass and fenced garden from the weeds, the thicket of brambles, a pin oak, a grove of spindly locusts, and some scrabbly maples that determine the start of the woods. My window doesn't offer much of a view, but what view there is simply is to a place where things change or where one thing becomes another. The edge is where things happen, and for that reason, it has always seemed a good place to work. From time to time in the winter, deer will amble out from the trees to see if the snowmelt by the garden has left anything edible and exposed, their step hesitant as they leave the cover of the woods. The aerial ballet of the gray squirrels is a constant, and on spring afternoons, it is not uncommon to watch a parade of wild turkeys marching with military purpose toward the brush Such small surprises are a good accomplice for any kind of work, and they remind me that I am adjacent to an unpredictable world. Witnessing them reminds me that things come and go unexpectedly.

Only once have I seen a coyote. It was a late winter afternoon when the ground was beginning to thaw, and the coyote was skirting the ragged line of the woods. It slipped out of the trees for a moment or so, an apparition, a quick brush of fur, a bit of smoke, a furtive, scruffy ghost that was gone as soon as it came. I caught little more than a glimpse of it, and I could barely have said what it was at all, much less tell whether it was a male or female, young or old. Much as I try to be attentive to what happens at the edge of these woods, it's hard to know how to record the particulars of a shadow.

But this tends to be how it is with the coyotes. The edge of the woods at the edge of the day at the edge of the season—this is the zone they have traditionally inhabited when they slip into the edge of one's vision, and a quick, illusory glimpse is the most they offer as they skirt the side of the road or lope along a field back into the swamp. In this part of the country, coyotes claim territories for themselves not much more than a square mile or two, and the woods and bit of meadow just behind the house is one such habitat. It's a clear-cut behind a stand of eastern white pines, and it's likely that the coyotes have a den in the rock ledges nearby. The coyote is monogamous, and a family pack may include a male, female, and pups from one or two litters. But the whereabouts of the den are conjecture only, no more certain than the perimeters of the area they hunt and patrol. The boundaries of their territory are as shadowy as the animals themselves.

Still, for all their obscurity, in the some seventy years since the coyote, *Canis latrans,* has taken up residence in this part of the country, it has managed to flourish. Larger than western coyotes and far more varied in color, these are the size of German shepherds, sometimes weighing up to fifty pounds. Speculation is that they migrated from the West, a decline in the gray wolf and mountain lion populations facilitating their advance. One group traveled north of the Great Lakes across Canada, while a second group traveled east through Ohio. The northern group bred with wolves during their travels, which is why the coyotes here now carry the DNA of wolves; and why, unlike their predecessors in the West, which eat only small animals, they are adept at hunting deer; and why, in less than a century, they have managed to adapt from grasslands and desert habitats to those of woodlands and forests.[1] But if eastern coyotes are such a hybrid, they do not live and hunt in hierarchical packs as wolves do, but rather, in small family groups.[2]

Roland Kays, now director of the Biodiversity Lab at the North Carolina Museum of Natural Sciences, was for years curator of mammals for the New York State Museum in Albany, and he has conducted extensive research on the eastern coyote. When I had visited him at the museum, he met me in the archives where skulls and pelts are cataloged and showed me a collection of skulls. The wolf's skull is significantly wider than that of the eastern coyote, and that of the eastern coyote slightly larger than that from Ohio. A wider skull, Kays explained, allows more room for muscles to work the jaws, thus attesting to the ability to hunt larger prey, in this case white-tailed deer instead of simply rabbits and small mammals. Further differences became evident when he took me to a locker full of pelts from eastern coyotes. Although the pelts of western coyotes tend to be a consistent dark, brownish gray, those of eastern coyotes are much richer and more diverse in color, ranging from pale cream and tawny to reddish blonds, smoky grays, and chocolate browns. Some are even pure black, Kays told me. It's a shuffling of genes, coyote, wolf, and dog.

But the coyotes' eastward migration of the last century altered breeding dynamics in another way as well: eastern coyotes also carry more DNA of domestic dogs than do their western counterparts. Western coyotes, which usually live in remote areas distant from any human population, breed with one another, thereby retaining a genetic distinctiveness. But the migration prompted coyotes to view domestic dogs, along with wolves, as mating opportunities. All of which begs the larger philosophical question: What defines a species? Is it reproductive isolation? Genetics? Geography?[3] What constitutes identity in the natural world is a realm of conjecture that seems not so distant from our own speculations about whatever obscure combination of factors—experience, ethnicity, inherited traits—collaborate to make us who we are.

Between twenty thousand and thirty thousand eastern coyotes

now live in New York State, numbers aided in part by the burgeoning human population. The mammals that thrive in suburbia—the raccoons that live behind the shed and forage in the garbage, the possums, mice, voles, rabbits, and deer—all make up the diet of the coyote. Coyotes are an invasive species but one that serves a purpose, Kays said. "Not all invasive species cause harm. These are filling the ecological role of the wolf, and they are now the top predator in the forests of the Northeast. While species recovery includes turkey, beaver, fishers, deer, it doesn't include cougars and wolves. So coyotes are filling that niche space." But as opportunistic feeders, coyotes are just as happy to eat fallen apples in the fall, raspberries, blueberries, birds' eggs, grasshoppers, flying squirrels, watermelon rinds, roadkill. In winter, they take down sick, old, and injured deer. They are adaptive, resourceful, intelligent, and the one I saw coming out of the woods that late winter afternoon was probably looking for a mouse or other rodent beneath the snow crust.

When I see what may be a coyote track, I am uncertain. I have been told that their prints are more compact than those of a dog and that only their two front claws will register. The tracks of dogs are larger and sloppier; dogs meander more, and all four of their claws tend to register. Coyotes walk with a greater sense of purpose. In winter, then, I sometimes try to decipher the coyote tracks. Sure enough, the tracks of my old yellow lab, Daisy, clearly register her habitual route along the garden fence, over to the woods, and back to the kitchen door. But as for the other tracks, the snow blows and melts, blurring whatever precision the original imprint may have had, and it is hard for me to be certain. In the end, I decide it probably makes sense to remain uncertain about the trail of something so elusive.

When I told Kays about the coyotes in the clear-cut and the one that had ventured to the edge of the woods, he confirmed their preference for such areas, telling me, "Prey species are more abundant in disturbed, younger forests where there is more going on on the ground and in edge areas where there is more sunlight and more wind than in the constancy of mature forests." The edge effect relates to the increase in both the variety and the density of animals and plants at the place where two habitats merge; species from both may converge there, at the edge of a lake, the edge of woods, the side of the meadow, all those small thresholds where one thing becomes another. Known in scientific parlance as "ecotone," this is the area where everything can change—soil content, temperature, humidity, light, vegetation, pollination—rendering it a place of vital and complex interaction. Such adjacencies, or borders, nurture biodiversity. White-tailed deer are drawn to the edge of the woods where they know they will find both the cover of trees and shrubs, bushes, grasses to graze on, and I think of Jeff telling my neighbor up the mountain where to position her birdhouse and how the presence of songbirds is known to be greater at the edge of woodlands and forests.

But if margins and edges are places of abundance, not all of them announce themselves explicitly. In developed areas, edges tend to be sudden, sharp, precise, a woodland ending abruptly at a vacant lot or suburban lawn; and in areas where there are many disparate uses of land, there can be an excess of edges. In nature, though, these distinctions in place can be more subtle, perhaps a shift in groundcover or a thinning of trees. The buffered areas provided by such gradual transitions allow for greater diversity and complexity in plant and animal life. And suddenly this idea that an edge need not be visible, apparent, or even remotely obvious seems like a fact worth paying attention to. Circumstances change without your ever noticing. Or sometimes the shift in conditions becomes evident only later. Human experience, it occurs to me, can benefit as well from an appreciation for such obscure edges, the way things are not always clear at the moment they happen, for example, or how words shift in meaning over time. There are times when a deficit of decisiveness allows for complexity of experience.

If it's rare to see coyotes, their sound is familiar. Their cry at night has nothing to do with edges but is, instead, a sound that takes up the full night sky, consuming and filling it at once. We live about a mile away from the local firehouse, and when the siren is set off, it is inevitably followed by their high-pitched cry. There is something in the shrill mechanical sound of the emergency siren that they recognize and that sets off their own instinct to call out. One winter afternoon several years ago, my husband was out on his snowshoes in the meadow up the hill from our house. He had just made his way through the pines along an old deer trail, and the stillness of the afternoon allowed him to imagine that he was alone. But as he stepped his way across the field, the firehouse siren began its wail, and just moments later the coyotes set up their own cry, seemingly within feet of where he stood. And I wonder what the coyotes are responding to, whether there is some instinctive awareness that the species they are answering to in this mournful and pointless exchange is foreign. Or perhaps it is the opposite; perhaps they are just answering to what thin strain of sound is familiar to them. Most likely, the siren's call is prompting them to delineate their territory, to broadcast their status, to announce themselves and the patch of land they have claimed. An eerie discourse between two quite different alarm systems, it remains an exchange that asks what makes *any* of us respond to the unknown, whether it is a search for the familiar or the curiosity about what is not. Sudden and spontaneous, it is a reciprocity that exists on its own terms.

Biologists sometimes use the sound of such sirens to track the coyote population. In remote areas of Idaho and Montana scientists have been using something called a Howlbox to track the wolf population. Equipped with a speaker-recorder system, it can be programmed to emit an electronic howl that then elicits a response from the gray wolves populating that region. Spectrogram technology that analyzes sound frequencies enables the scientists to distinguish among the assorted responses, thus helping them to count the number of wolves making them and, ultimately, assist in the long-term management of the wolf population.

Coyotes do not harmonize as wolves do; their pitch is higher, and the range of sounds a small family group makes is more diverse. Their vocalization includes eleven basic sounds that range from yips and barks to growls and whines, whimpers, low cries. But such words, of course, bear little relation to those declensions of sound that actually emerge from the night woods. A fugue, I think. Or an aria. But those words are not right either. And the function of these sounds is as diverse: a call to mate, a summons to the young, to announce territory, to establish dominance, to welcome, to warn.

Listening to them, I try to be more alert, more attentive to the gradations of sound. Places reveal themselves through sound, the peepers on an April morning, the dry leaves blowing across grass in November, the thunderous fracture of ice on the river in January. As any birder will attest, landscape can reveal itself through sound.

Noting the lack of domestic animals at his cabin, Thoreau heard only "squirrels on the roof and under the floor, a whippoorwill on the ridge pole, a blue-jay screaming beneath the window, a hare or woodchuck under the house, a screech-owl or a cat-owl behind it, a flock of wild geese or a laughing loon on the pond, and a fox to bark in the night."[4] If sound situates us in place, perhaps it is because hearing has a different, more intimate relationship to memory than sight. Sounds converge in some sensory region of the brain, their association more primal, their alliance a prompt to the imagination. This is why you can remember words months and years after they have been said, recapture their tone and cadence precisely, tune into their frequency exactly, and find new meaning in them; or the way listening to a long-forgotten piece of music can bring an earlier moment of your life into full bloom. And certainly the sounds of the natural world trigger ancient fears and pleasures. The association between listening and landscape may remain so vital today simply because our reptile brains retain some vestigial knowledge our ancestors acquired on the savannah, knowing how danger so often arrives through sound. It is the quiet of birds that signals menace and disorder, while their song delights us exactly because it delivers us to reassurance.[5]

Bernie Krause, acoustician and author of *The Great Animal Orchestra,* maps landscape through sound, and his niche hypothesis posits that the diversity of sound in a particular habitat indicates its health. Krause maintains that the vocalizations of the bird and animal kingdoms have evolved over time, the aural signal of each inhabiting its own piece of sonic real estate in the greater soundscape, and that this vast symphony of living organisms is precisely calibrated to ensure the well-being of each of its players. Whether it is a mating call, warning signal, territorial claim, or some screech of distress and discomfort, a creature's sounds help to ensure its survival. Although they remain a mystery, the nuances of the coyotes' cry make me think this might be true. I can't quite say where a bark becomes a whine or a yip a howl, any more than I can say where fear and desire converge or where a welcome may intersect with a warning. The variations of sound, at once fierce and plaintive, simply blend to my ear, their conversations rippling freely from flippant greetings to claims of authority, with only a few notes distinguishing one from the other.

When I had put these questions to Kays, he simply said, "It's hard to judge sounds in the wild." I take him at his word. Perhaps it is only natural for us to want to understand, comprehend, decipher, catalog these cries, to match each with the intent it is expressing. We want exactitude and answers, and it's what sends us running to our iPhones and Androids every time some twitch of curiosity is triggered in our brains. But interpreting the coyote's call is not a matter of uncertainty; it is a matter of not having a clue. The sound occurs in a chamber of the inexplicable, and Kays's blunt reply reminded me that the unanswerable also has a place in our lives.

Besides, there is something in the frequency of the calls that allows the coyotes to seem more distant than they really are. Listening to them, one is simply reminded of that fugitive equation between sound and intent; miss an inflection and you can miss meaning altogether. With their repetitions and echoes, the cry of two or three coyotes will sound like twenty, announcing their presence with a distant finality. This trick they have of seeming to double, triple, quadruple in number as soon as they set up their cry is another of their vocalization skills. Yet while the cry is collective, it manages to remain a soundtrack to solitude, and this ambiguity seems essential to their message—you are alone; you are not alone. Somehow, their choir manages to say both of these things.

In the twenty-first century, we are unlikely to find in the animal realm those totems to which our predecessors so often looked. It tends to be the science of animals rather than their spirit world that elicits our interest. But I wonder if the coyote is the exception to this rule; coyotes and humans have a long tradition of meeting in the territory of myth and fable, and the impulse to align ourselves with them—or with their cunning, wile, independence, feral solitude or any other of the characteristics we attribute to them—may not have vanished entirely. A 2008 study of how coyotes and humans interacted in rural areas just south of here found that although most residents had had some experience with coyotes—seeing them in the yard or listening to their calls—they were likely to think they were alone in such encounters and that their neighbors had little knowledge or awareness of coyotes.[6] It seems we like to imagine that our own associations with this animal are private, singular occurrences. For reasons not entirely clear, the solitary persona of the coyote manages to infuse our own experience of them.

The coyotes here tend to keep to their territory up the mountain, but south of here where it is less rural, they have begun to adapt to the human population. Invisibility is no longer their trademark, and from time to time they are found loping across suburban golf courses or, having found their way to Manhattan, sauntering through Central Park or sprinting through Tribeca. Although this is partly due to habitat loss and increased familiarity with humans, there is some speculation, too, that the eastern coyotes' hybridization with domestic dogs has helped facilitate their growing ease with humans.[7] Every spring and summer, stories abound of household pets attacked by them and children playing in yards frightened by their sudden appearance. The coyotes' comfort around us has become a hazard to them, and we are advised to cultivate their fear, to bang on kitchen pans to frighten them if they come to the yard, throw rocks or sticks and otherwise chase them away. It is the same argument that advocates hunting and trapping them, a more extreme and effective means of keeping their fear of us intact—which, in the end, may be what protects them most.

A 2006 study of coyotes in Westchester County used citizen science both to map their habitats and to document their interaction with humans.[8] Structured around a voluntary class assignment, the survey asked schoolkids to document whether and where coyotes had been seen or heard. Not surprisingly, families who lived close to woods and grasslands reported hearing and seeing them more than those living in more developed areas. Mark Weckel is a conservation biologist at the Mianus River Gorge Preserve in Bedford, New York, who managed the study, and when I reached him by telephone to ask him about the perceptions among those families that participated, he told me that in the northern part of the county, where it is more rural, properties are larger, and the population is more accustomed to wild animals, residents are less likely to see the coyote population as a problem. But in the southern part of the county, which is more urban and suburban—and where there is less familiarity with coyotes—fear often persists, and the coyote population continues to disturb residents.

It is Weckel's hope that studies such as his will not only enable wildlife professionals to better manage the coyote population but also help homeowners learn how to accommodate this emerging presence in the suburban landscape. As the report concludes,

> The future of coyotes in suburban areas such as Westchester County depends not only on a scientific understanding of urban coyote ecology but also on stakeholders' willingness to share their backyards with a top predator. Our approach of using citizen-generated data serves to elevate the profile of the novel predator in the community, a prerequisite for managing potential human-wildlife conflict, and also permits stakeholders to quantify their own risk. Science has been criticized for being opaque and therefore can become underutilized by the public it is meant to serve . . . . Citizen science seeks to engage the stakeholder in the process of knowledge building on the premise that, by doing so, the stakeholder becomes more actively engaged in the environmental problem at hand.[9]

But I know we don't share the backyard so much as we share its edge. There are times, after all, when sharing is misguided, occasions on which the healthy liaison is the one marked by fear and abiding distrust; when stewardship is a matter of maintaining hostilities, when admiration is best practiced with detachment, and when caring for something or someone is a matter of measured distance. And while I try to think of myself as a stakeholder, I know that what I really am is just a common listener. And that the coyotes' remote orchestra is a reminder of how sound waves can be the threads that connect us to place.

When I hear them at night, it is hard to know whether I am already awake or if their cry has roused me, triggering some ancient association of sound and fear. At that hour of the night, the particulars of the sequence don't seem important. The predictable rhythm of the siren and the unpredictable song of the coyote make up their own dialogue, and listening to the eerie duet of the animal and mechanical world, I am certain that I am hearing an authentic soundtrack not only for life in this valley but for this time of improvised transactions between the natural and manmade worlds. And if my mind then wanders to work or health or money or the well-being of my children or any of the other concerns that visit us so effortlessly at that hour, the distant symphony inevitably braids a more primal anxiety into these ordinary worries. All I can think as I try to fall back into sleep is how fear is essential to who we are and what we do.

"Wilderness settles peace on the soul because it needs no help," writes Edward O. Wilson. "It is beyond human contrivance. Wilderness is a metaphor of unlimited opportunity, rising from a tribal memory of a time when humanity spread across the world, valley to valley, island to island, godstruck, firm in the belief that virgin land went on forever past the horizon."[10] Yet we have arrived now at that strictly modern paradox that wildness is something that is tended to, arranged, nurtured, some quality that must be cultivated, managed, engineered. Most of us have fully lost the meaning of wildness, its mystery, its fullness, and its unascertainable character gone from our lives. If wilderness areas exist now only because they have been carefully planned and preserved, it seems fitting that the enigmatic persona of the coyote serves as some kind of brand mascot for this new acreage, its cry reaching us as it does simply because it is a primitive reminder, an acoustical souvenir of this ancient sanctuary.

And I think of the coyote skirting the woods on that late winter afternoon. If I was unable to determine its age or its gender, if I did not know what it was seeking, its intent or its destination, what I knew perfectly well was that, for an instant, I was observing something both foreign and familiar. Its size, its proportions, its ragged gray coat, the twitch of its ear all identified it as a canine and signaled its genetic proximity to a familiar domestic animal. Yet it was more feral in its bearing, quicker, more distant, and indifferent to me. Freud defined the uncanny as something that is known yet manages to retain a quality of the unfamiliar, and there is something of that here. As disquieting as a wild animal may be, it is eerier still when it retains some component of the known, and it is especially unsettling to us when the known and the unknown collaborate this way in a single creature, when we are confronted with a convergence of two natures we believe to be contrary. Maybe it is a matter of our own shuffled impulses that makes us so responsive to the shuffled genes of the coyote. I wonder if our fascination with the coyote has to do with this convergence of the domestic and the wild that beguiles and repulses us at once. And if the reason that the coyote's howl manages to slip beneath our skin with such ease is on account of this small edge habitat we all carry within ourselves.

7

Herring into the Brook

What human being who has even a passing familiarity with desire or expectation or loss or regret could possibly pass up a job that requires standing at the edge of a bridge for fifteen minutes twice each week simply watching the water streaming beneath?

Wappinger Creek is the longest creek in Dutchess County,

starting at Thompson Pond at the foot of Stissing Mountain in the north and running some forty-one miles south where it drains into the Hudson River. One of its many tributaries, Hunter's Brook is a small freshwater stream that meets the creek not far from the river. A small bridge spans the brook at its mouth where it meets the creek, and that line where the dark, clear water of the brook meets the cloudier, brackish water of the creek is usually visible and precise. That border is always shifting, though, depending on rainfall or how much freshwater is running down from the brook or how the tide from the Hudson has flowed up the creek and rearranged its silt. Perhaps the oddest thing about standing on the bridge at Hunter's Brook is that it is a place where you can actually see the water flowing in two separate directions simultaneously. A film of spring pollen, floating leaves, or bit of light foam from storm overflow might be washing downstream, yet at the same time, the light can catch the ripples of water as the tide washes up the creek.

That enigma runs in the current here makes it an appropriate site to watch for the herring migration, because the herring themselves are a puzzle. Herring are anadromous, which means they spend much of their life at sea before returning to freshwater to spawn. Along the Atlantic coast, from Florida to Newfoundland, they return by instinct each spring to the tributaries in which they were born. In recent years, though, their numbers have diminished, possibly because of the proliferation of dams, a growing population of predators, overfishing, compromised water quality, or habitat loss. In an effort to establish some baseline understanding of the changing migration patterns, the New York State Department of Environmental Conservation's Hudson River Estuary Program and Hudson River Fisheries Unit, along with the Student Conservation

Association, instituted the Volunteer River Herring Monitoring Program in 2008 to help gather data about both the herrings' spawning habits as they arrive in the river's tributaries and about the environmental factors that may influence them. In 2011, 274 volunteers monitored twelve sites from the Saw Mill River in the south to the Poesten Kill in the north.

It is one of those endeavors that require little more than attentiveness. The herring do not need to be collected, measured, or weighed but simply observed and, if possible, counted, and there is nothing in the way of technology required unless one counts polarized sunglasses and a thermometer. Volunteers are asked to spend fifteen minutes twice a week, from early April until the end of May, at one of twelve Hudson River tributaries to watch for the arrival of the herring, noting such facts as tide stage, water temperature, air temperature, cloud cover, and precipitation. The data sheets include space for "other information," which could range from the double-crested cormorants, the muskrats, and the barn swallows to the tired grievances of the anglers on the shore; it can be the place to note what is seen when that animal or plant you are looking for does not happen to materialize. Or perhaps in this rigorous catalog of fact, a little bit of blank space for the intuited. But what this exercise in observation is about most is whether the herring are here or not. Absence or presence. In the digital age, when visual information seems to assault us in a relentless blizzard of imagery, graphics, and signage, trying to see one thing clearly seems a relevant exercise.

Because the lower Hudson River is an estuary with its own cycle of tides, the flow of water in its tributaries is also tidal for a short distance. And at high tide on an early April morning in 2011, the channel of the creek was a murky green; where a gravel bar lay beneath the surface was an even muddier hue. Three bodies of water are near one another here, and how their currents converge and their banks are shaped by changing combinations of tide, current, rainfall, and snowmelt is never quite the same; the architecture of the river is under constant reconstruction. A few days later, as the tide ebbed, the bar seemed to become fully visible within minutes. But by then we were four days deeper into the spring. The acid green of the skunk cabbage sprouting at the far edge of the creek, a vivid crack of color in the dusty brown of the woods behind it, brazenly announced a new frontline for the season.

Cascades of forsythia had begun to appear in the gardens and yards of the houses I passed on my way to the creek, and once I got there, an angler tossing out his line told me that the alewives, the first wave of herring, usually arrive when forsythia is in full bloom. Phenology, the study of how things unfold in nature on regular and predictable cycles, is about the seasonal order of things, and in recent years, it has become a valuable tool in the study of climate change; when alignments are disrupted, the food chain can fray and species suffer. But one of the things I love about phenology is how it can reveal coincidences of seasonal events and align such arbitrary partners—such as this unexpected alliance of forsythia and alewife. A few weeks later in the season, flowering lilacs traditionally signal the arrival of the blueback herring. And shadbush is the customary name for *Amelanchier arborea,* a common shrub whose white blossoms tend to appear at about the time the shad run up the Hudson River. I am certain that just as our sense of measure is derived from such timing and simultaneity, so, too, does it thrive on such surprise adjacencies. But if the brook was running high with snowmelt and color beginning to settle in to the woods, the herring remained out of sight. The air temperature was barely in the upper forties, and the water temperature remained below fifty-one degrees Fahrenheit, the presumed threshold spawning temperature for herring.

Seasons never arrive definitively, and April in the Hudson Valley is a time of disconnect. On April 11, after a strong rain, the woods remained leafless and brown, though green sprays of multiflora on the verge of being in full leaf began to appear like stray bits of green cloud that had drifted into the woods. The lawns were emerald green, and fluorescent tufts of dandelion, their brightness a little eerie after the long winter, had appeared overnight on the side of the road. Trees were in bud, though not a leaf was in sight. Still, the thermometer had hit the mideighties and the air was sticky with a breath of summer for the first time in six months. What could only be thunderclouds were moving in from the west. Not surprisingly, the temperature of the water had risen accordingly into the low sixties. I didn't get to the brook until early evening when its waters, at flood tide, were flowing languorously into Wappinger Creek.

Beneath the surface of the water a massive carp drifted listlessly along the river bottom, and moments later, a muskrat slid along the concrete embankment of the bridge. Both are creatures of weight and substance, but moments later, something rippled beneath the water's surface, a shadow crossing a stone on the riverbed, a flash or iridescence, a quickening in the water. Then I saw them, five, fifteen, twenty slender fish, the backs shining dark, stomachs a more translucent green, as the complexion of the water shifted to take on a manic shimmer, the herring moving in flashes and curves, bending and twisting in a current all their own. I saw now that the distinguishing feature was not shape or size as much as the manner of

Herring into the Brook

movement and speed. Herring are ten to fourteen inches in length, they have a small dorsal fin, a forked tail, a spot behind the gills, but none of these are what one notices half as much as the rhythm and pattern of the watery caravan as it circles and swirls beneath the creek's surface.

It was dusk, with the visibility of the water dimming, and I couldn't make out the spot behind the herrings' gills any more than I could tell whether these were alewives or bluebacks. But the shift and shimmer and twist and curve spoke to their certain arrival. Scientists have conjectured that the migrating herring may use the flood tide to give them an extra lift on their way into the freshwater streams, and it was hard now not to imagine these cavorting schools of herring were taking some pleasure in their own rite of spring as they rode the river's tide into the brook at the end of the first hot afternoon of the year. The data sheets offer options beginning with zero, then go to 1 to 10; 11 to 100; 101 to 1,000; and finally 1,000. There were more than ten, I knew, but the quickness of their movements, the speed with which they came and went, left questions about whether I was watching the same school come in, circling and doubling around on itself or if one school after another was swimming into the brook. But if my observations were simply about absence and presence, it was clear that the brook had gone from one to the other.

Yet when I went to the creek in the late morning a week later, I saw nothing. Training my eye to look into the moving water asked for a shift in focus. It is only natural for the eye to follow the direction of the brook's downstream flow into Wappinger Creek. But the herring were swimming upstream into the tributary, and it takes a strange optical effort to find something that is both beneath the surface and swimming against the current—a manner of squinting that is as much cerebral as optical. Even managing that, the flood tide and heavy rain of the previous night had churned up the silt, the creek was roiling with mud, and visibility was poor. A thin film of bubbles, possibly some residue from a storm overflow, in the brook's current further obscured things. High, fast water creates its own energy and rhythm, and the swallows drifting and darting above the creek added to that sense of animation. Possibly the downstream torrent discouraged the herring from making the trip up the tributary; they may have been out in the river, waiting for the waters to calm and for more assistance from the tide to help them get upstream. Or perhaps the flood of storm water had prevented them from reaching the tributary. Nothing in the dark water offered answers.

In her essay "Seeing," Annie Dillard writes that "it's all a matter of keeping my eyes open. Nature is like one of those line drawings of a tree that are puzzles for children: Can you find hidden in the leaves a duck, a house, a boy, a bucket, a zebra, a boot? Specialists can find the most incredibly well-hidden things."[1] Looking into the water now, I know that such watchfulness is something to aspire to. I think back to the opaque vernal pools I had seen last April. The creek today is just as unfathomable. This, too, is water that gives up nothing. If there are herring beneath the surface, curving and twisting in their own fleet upstream, I can't see them. "The lover can see, and the knowledgeable," Dillard has also written.[2] And though I know she is right, I also imagine that it may now sometimes take both loving *and* knowing to observe what is important.

A tide chart is an innately soothing document; something in its columns of numbers, its neat arrangement of figures, speaks to the way we have managed to find some order in the watery chaos of ocean currents. Yet from time to time, its format can be confusing. Most often this has to do with spacing. The date, the height of the water level, the time, and whether the tide is high or low—all of this is straightforward enough, but the columns of information are sometimes aligned so that it is easy to confuse the status of the tide and the hour of the day with which it corresponds. The chart from the National Oceanographic and Atmospheric Administration I have been using has been laid out with these misleading margins and perplexing configurations, and I find I have to check, then double-check my observations. It is an example, I suspect, of the document reflecting the nature of the experience itself. It is easy to be careless about the space between things. At Hunter's Brook on a late April morning, the tide is just coming in, shifting that space between the creek and its bank. Where the brook and the creek meet is a moving line as well, and the rapid darting to and fro of the herring schools makes it difficult to distinguish their numbers. If it requires a little extra attention now to read the chart or determine whether it is a flood or ebb tide, such uncertainties only speak to the ever-shifting space between what we imagine, what we see, what we can do. I look back into the water. A cloud passes over the sun, and suddenly the water is dark and impenetrable again. "Accuracy of observation is the equivalent of accuracy of thinking," Wallace Stevens wrote in *Adagia,* and it occurs to me that these words are strong evidence that poetic conjecture and scientific inquiry are more similar than we like to think.

One lesson this tutorial on attentiveness has offered is that what is seen is not so much about what happens to be there as it is about the conditions of visibility. These vary. The tide may be low, the sunlight bright, the water shallow, but still, there is nothing. On a gray day, when the tide is coming in, the obscurity of the brook seems only to intensify with each passing minute, the water getting deeper, darker, more opaque. It may be water, but it has nothing to do with transparency. How long does it take to see something, I wonder. Fifteen minutes seems as good an answer as any.

At the boat launch, a couple of men were getting ready for an afternoon of fishing. Perch and crappies, they told me, easy pan fish. I asked them if they had seen the herring, and the younger one shook his head. "I've been fishing here my whole life," he told me, "since I was knee high to a grape, and there are fewer and fewer each year." His father, a man well into his seventies, added, "They used to be so thick in the water you could practically walk across them." The exchange, though brief, was an evocative illustration of something called the shifting baseline syndrome, which is the continuing adjustment in accepted norms for ecological conditions. Changing circumstances cause succeeding generations to experience the natural world with shifting expectations—whether it has to do with the diminishment of a particular species, changing amounts of rain or snowfall, or adjustments in temperature. The direction of the shift is generally downward: if the elderly fisherman's standard came from a recollection of abundance, his son's was derived from witnessing scarcity. I was reminded of the carpet of water chestnuts that Nancy and I had tried to thin and how the presence of those weeds now signals a new norm.

I wondered at that curious alliance between observation and memory. In the weeks after the break-in at our house, I had scrutinized photographic lineups of suspects. I remembered the winter morning and seeing the burglars in my driveway. Weeks later, though, it was next to impossible to align the photographs in front of me with those images in my memory. A lock of hair, the line of a cheekbone, an expression might intersect, but beyond those, nothing. Witness accounts are fallible, and memory, susceptible. About a third of the seventy-five thousand witness identifications that are made each year are incorrect: we are prone to distraction; we overlook details; we are open to suggestion. I looked back at the shadows in the water. Even if you see something clearly with your own eyes, you can get it all wrong.

Yet it is a paradox of human perception that forgetting is part of knowing. My father, a journalist and war correspondent, knew something about documenting facts, taking notes, and getting the details right. Yet he told me more than once that the human brain was a machine for forgetting, and that what was most miraculous about human thought was how much information and experience was naturally discarded. *Document* and *keep* are vital, but so, too, is *erase*. I think of this now and wonder at how the human imagination chooses what to save and what to relinquish. On this April afternoon the way such selections are made seems as fugitive as the quick water and bright shine of the migrating fish themselves.

I walked back to the bridge and looked into the water for a while longer. Although the letters of his name had faded, Danny Moffat left his signature with a marker on the galvanized metal railing of the bridge in January 2008, and over the weeks, I had taken to thinking of my interludes at the bridge as appointments with Danny Moffat, my little mascot in the presence and absence project. I wondered at that impulse we so often have to leave a trace of ourselves, to leave that kind of signature or tagging that seems to inevitably show up at places where we *wait*. And I wonder if he has ever read Borges and come across the line, "What man of us has never felt, walking through the twilight or writing down a date from his past, that he has lost something infinite?"[3] Or maybe Danny Moffat already knew that. His elusive attendance here seemed somehow suited to this exercise of watching for a migration that you cannot be certain will take place.

A few days later, I was almost certain of the herrings' return. There hadn't been any rain, the water in the creek had receded, it was flood tide, and the air was warm. Sunlight and a clear sky gave the water a brilliant clarity. Dogwoods and magnolias were beginning to flower, the lacy threads of the willows were greening, and the air had taken on the full softness of spring. Yet again, nothing. There was something mesmerizing in the combination of repetition and surprise, something about the rhythm of the familiar and the unexpected so intrinsic to these visits, and when I mention this to my friend Polly, who sometimes joins me at the bridge, she said, "But it's the waiting, too. That's what makes it so meditative. You just keep coming back to the same place. You know something is going to happen, but you don't know what it is."

I stared into the water, almost wishing them into existence, and it almost happened in this little factory of misconceptions. Large, brown oak leaves from an earlier season floated downstream, creating elusive shadows on the bed below. The sun passed through the scalloped surface of the water and rippled on the bed, playing with the shadows of the rocks. What was below the surface was a choreography of illusions, one that could easily accommodate the fugitive iridescence of the herrings themselves. The anglers I see at the river have speculated that the hook shape of the gravel bar may act as a barrier, discouraging the herring from making the turn into the brook. And one afternoon earlier that week, a fisherman out with his two young sons had suggested that as striper fishing has become more popular, the herring, used as bait, have been depleted. Two new explanations for absence. Whether there is truth to these is anyone's guess. What I do know is how inventive we can be when it comes to explaining to ourselves why something we want to be there isn't.

For scientists, uncertainty has a precise meaning that has to do with variability in data. It is the unknown in a data set, and although it seems odd that uncertainty can be measured with such accuracy, it is often measured in a percentage. Likewise, the words "confidence" and "significant" have exact definitions. For most people, the term "confidence" is a matter of personal judgment. For scientists, however, it has statistical significance; confidence limits in data reflect the similarity of all of the numbers in the data set, and conclusions are made only when the data points are 95 percent consistent. Likewise, something is "significant" in science only if it meets precise mathematical attributions—or when the confidence limit of the data is 95 percent or more. In science, nothing is "significant" unless a statistical analysis of the data reveals that specific value.[4]

On this April morning, it remains a mystery to me that doubt can be measured so fastidiously. I imagined these shining ribbons of silver streaming beneath me are the migration. Or not. If looking for herring is about separating presence from absence, it is also about separating observation from apparition. In the shifting light of the water I saw now that looking for the herring is one of those exercises of trying to locate the elusive in the evanescent, maybe some cousin to a double negative. And maybe that means they are there after all, though I knew such word games didn't go far here.

I thought of a drawing class my mother once took after my sister and I had gone off to college. She had found time to take on other things, or maybe, like her daughter decades later, she, too, simply wanted to teach herself to see. In the class she was instructed to spend four hours drawing an orange, six hours drawing a fold of cloth, twelve hours sketching the sculpted head of a man. She taught herself to separate shape, space, shadow, texture, light. She spent hours drawing a leaf of lettuce. She took her sketch pad home and showed that page to my father and asked him to tell her what it was. He stared at it long and hard and finally turned to her and said, "I think it's a brooch." It became a running joke in our family, some riff on marital differences, some bit of proof that despite all that practice and discipline, in even the most enduring marriages, what is to one person something you put in a salad was a piece of jewelry to another.

I knew by now that my time at the creek each week—which is to say, 30 minutes inside of 10,080 minutes—was time to see what happened to be there. I knew that in the remaining 10,050 minutes I would see what I wanted to see or hoped to see or expected to see, all those small strategies we devise for ourselves so we can have what we want to have, believe what we want to believe; when we confuse presence with absence and absence with presence. These 30 minutes were a controlled exercise to see what *was* there. And then I thought of what I wanted to see and of the two men out in their skiff, fishing, and of Danny Moffat standing at this bridge and writing his name for whatever reason he had and of the herring that may be waiting out in the river for the water to recede or the weather to warm or the tide to rise, and I wondered, what if any of us got *what* we wanted *when* we wanted it?

When the herring finally returned, it was not in the dense rush I had witnessed three weeks earlier. Water temperatures had risen, and high numbers of herring had been spotted in the Quassaic Creek, the Fishkill Creek, the Saw Kill Creek, the Hannacroix Creek, Black Creek, and Wappinger Creek. Bluebacks as well as alewives were appearing, and sightings were only expected to rise. At Hunter's Brook, though, they appeared in sparse, languid schools on a warm breezy day in early May. The water was quieter, too. The spring rains had subsided, and it was ebb tide. When clouds passed over the sun, the surface of the creek darkened, and seeing *anything* beneath the surface was nearly impossible. Two anglers out for bass swore not to have seen them at all, and I nearly missed them as well. In their first run weeks earlier, the herring had arrived cavorting in an obvious stream, but their arrangement in the second wave of arrival was more subdued. Absence may have gone to presence, yet they remained few and intermittent. Such a rhythm of egress is, of course, the reversal of what you would have expected, but that was likely the point of being out there with the data sheets.

My last morning at the brook was hot and sultry, feeling more like July than the end of May. At flood tide, the sun was bright, but a strong rain a few hours earlier had again muddied both brook and creek. It was impossible to see what was happening beneath the water's surface, though a violet tail dragonfly skittered just above it. Petals from flowering locusts floated downstream, along with a dusting of pollen from a nearby cottonwood. Though I knew it was not the objective of the monitoring project, I had found that it was often impossible *not* to time my outings to the brook with those times I expected to see the herring—bright sunlight, warm water, flood tide. If looking for something, it's only human nature to try to imagine the conditions under which you might best find it.

I learned later that over the season, the volunteers had made a total of 711 trips to the twelve sites, recording 179 sightings. Eighty-nine percent of those sightings had occurred when the water temperature was above fifty degrees Fahrenheit. And at sites below the head of the tide, the chance of seeing herring during an incoming tide increased by 7 percent. These are the known facts. Still, I wondered whether my colleagues in the absence and presence project had concluded, as I have, that herring are reluctant to align themselves with expectation. I knew by then that clarity does not always reveal. Herring are more likely to be seen when water clarity is "fair" rather than when it is "excellent"; they may be more likely to migrate upstream when visibility is compromised, allowing them to take cover in shadier areas where they will be less apparent to predators.

But today, there was no telling whether they were here. Out in the channel of the creek, the carp were slapping the water, their dance creating random rings of ripples. I looked again across the water. Perhaps such surveillance, seeing the world in front of our eyes exactly as it exists, is an act of the imagination. The naturalist John Burroughs wrote, "The habit of observation is the habit of clear and decisive gazing; not by a first casual glance but by a steady deliberate aim of the eye are the rare and characteristic things discovered. You must look intently and hold your eye firmly to the spot, to see more than do the rank and file of mankind."[5]

His instructions on watching, published more than 130 years

ago, clearly come from another time. We are inclined now to put a greater value on being watched than on watching. But in my late fifties, I often experience a veil of invisibility, a kind of indifference that is not entirely unwelcome. On this late spring afternoon, I know that the active end of observation is a more hopeful place to be, and that a steady, deliberate aim of the eye is something to be grateful for. I know this; my confidence limits on this are well above 95 percent. Sometimes it is enough simply to record presence or absence and the conditions under which they occur. A swallow skimmed along the surface of the creek, a great blue heron took flight from the opposite bank, the level of the water dropped imperceptibly as the tide receded. At the edge of water, any water at all, you would have to be a fool not to want to be the one who is doing the noticing.

And it occurred to me then that this is a place I may return to. I may not know what propels the herring into this tributary or whether it is the gravel bar or the water temperature or some predator that keeps them out. But I do know that these fifteen-minute increments at the bridge over Hunter's Brook can serve as a template for attentiveness. It is a random but reasonable interval, neither too little nor too much, and I tried to anticipate those moments in the future, when a decision is made or a diagnosis delivered, when some piece of essential information arrives suddenly, when I might be able to find my way back here, when I might *imagine* that I am standing here at this bridge. For fifteen minutes exactly, I will try to distinguish between the water and the light, the shadows and the fish, to separate fact from apparition, to consider the conditions of visibility, to recognize those baselines that may be shifting, and, above all, to see what is there.

8

Loosestrife in the Marsh

When I was a teenager, I would sometimes have tea with an older woman who was a friend of my mother's. Mrs. Bacon lived alone, and my mother was of the mind that if a seventeen-year-old-girl were to discuss books with a seventy-five-year-old woman, both would be the better for it. Now, decades later, I am able to recall little of the books or conversation, but I remember clearly one afternoon, after our talk and our cup of tea, Mrs. Bacon took me to the wide window overlooking the Dutchess County meadow. It was late summer, and the field of rye grass and timothy was alive with color, chicory, Queen Anne's lace, joe-pye weed. She drew back the damask curtain and waved her hand out toward the meadow, then turned to me and said, "Oh, my dear, you will never believe what I saw this morning. I stood here and looked out the window, and there, loping across the field, was a Russian bobcat. How it got here all the way from Russia I shall never know, but get here it did!"

To this day, when I drive by that field, I keep my eyes open for a Russian bobcat. Today, though, it is a different landscape, and what is there now speaks to changes in land use in this part of the Northeast. In the 1970s, the meadow was used by a local farmer as pasture for his dairy cows, but it has since been parceled into building lots with several small ranch houses whose owners have planted ornamental fruit trees and small spruces in their yards. Willows and sycamores have grown in along the creek, and a large pond shimmers in an area originally excavated as a gravel mine. And while I have yet to spot the bobcat loping through the tall grass, a different exotic has taken up residence there. Where the land is wet, where the marsh meets the pond, vibrant stands of purple loosestrife proliferate.

Hints of the loosestrife first appear in late July or early August, a stalk here, a slight stand there. Within days, the tall spikes have spread, a trench along the road is lined with them, a depression in a field is suddenly painted with a magenta brush, a marsh along the river is suddenly florid. In such a way, the contours of the landscape are reaffirmed. Wherever there is a bit of moisture, a swamp, a piece of wetland, the bank of a pond, the edge of a lake or a river—all of these are suddenly transformed by vibrant purple wash. By the end of the month, the bogs across the road have become a thicket of these magenta spikes, some nearly ten feet tall and often edging out the cattail and tussock sedge. This is sometimes known as a mono-typic stand, meaning it is composed of a single species, but on a hazy August morning it looks more like some refugee cloud from daybreak has just decided to linger on the ground.

Its color that of bishops, royalty, and imperial entrances, the loosestrife arrives with a like sense of authority, and even the flowers have a name that implies some kind of botanical regency: "inflorescence" describes the clustered arrangement of flowers on an axis, in this case, five- to six-petaled flowers on a stem that can be up to sixteen inches in length. But its evocation of rank is due to more than its color, because its numbers, too, are spectacular. Each stalk of loosestrife can bear as many as fifty smaller stems that produce between two and three million seeds each year that can be spread quickly by wind and water. Beauty and tenacity have always been compelling partners, and nowhere do they make a better showing than in the Hudson Valley wetlands in late summer.

These perennials are thought to have arrived in the United States about two hundred years ago as an ornamental and medicinal plant for the treatment of diarrhea, dysentery, ulcers, and assorted sores. Since then, they have developed a reputation for botanical ruthlessness. Like many other nonnative plants, they meet few natural enemies when they arrive on foreign soil and are able to grow quickly. With the ability to spread rapidly, loosestrife appeared to be edging out native plants, strangling the cattails, displacing wildlife, and generally diminishing the biodiversity of those places where it took root. Populations of waterfowl such as marsh wrens, American bitterns, and black terns were thought to have been compromised and, in some cases, threatened with extinction. A 1987 paper addressing the control of purple loosestrife noted its disastrous impact on native vegetation, the threat of declining species of vertebrates whose breeding habitats were compromised by the plant, and a grave diminishment of waterfowl.[1] Unlike the elusive bobcat that slipped into sight, or into a woman's imagination at least, for an instant from the haze of an early summer morning, the loosestrife for decades has been more present, more visible, more persistent.

But all that said, perspectives about the loosestrife have shifted in recent years. Since the mid-1990s, its growth in New York State has been reduced by a biological control—leaf-eating beetles that nibble on the leaves and stems of young plants in spring while females lay their eggs on the stems or ground nearby. The beetles do not eradicate the loosestrife so much as diminish it, and it is not unusual now to see wetlands where its tall spikes mingle more collegially with other marsh weeds and flowers.

Erik Kiviat, executive director of Hudsonia at Bard College, Annandale, New York, has spent a good part of his professional life studying purple loosestrife and its place in the regional ecology. When I visited him at his field office at Bard, he told me he had grown up in the Hudson Valley in the 1960s and remembers watching the late summer bloom of loosestrife at the pond of his childhood home and how the red-winged blackbirds nested in it. Slight and soft-spoken, he speaks in the quiet way that people with conviction often do: "What I saw, then, was that the red-winged blackbirds actually preferred it to the cattail. Loosestrife is more stable and robust. The nests sustained more damage when they were built in the cattails. Also, the loosestrife stands had lots of insects, moths, butterflies." Years later, in the early 1970s, as a young scientist doing fieldwork in Tivoli Bays in northern Dutchess County, he looked more closely. Noting that "research on weeds just tends to be on how bad they are," Kiviat became interested in everything he could see about the plant and its unexpected—and largely unexamined—function as a habitat.

Is loosestrife the aggressor we assume it to be, or is it a competitor in a world of ever-changing ecologies? Is what appears to be a predatory invasion, in fact, more of a tenacious adaptation to changing conditions such as a warmer climate, a disrupted habitat, or, in the case of plants that thrive in moisture, a result of changing drainage patterns? Does it signal environmental devastation, or can it be a catalyst for a new brand of diversity? And has it been our own poor land-use decisions that have done the most to cultivate such species?

Kiviat found that although there may be some negative impact, "it was not as much as people think. Purple loosestrife is a top-notch honey plant, a nectar source when other sources have dried out. It is pretty drought resistant, and its nectar production is prodigious—lots of bumblebees, though lately fewer honeybees." In the late 1990s, Kiviat engaged volunteers to observe stands of purple loosestrife not only in the Hudson Valley but in other areas of the Northeast. The diverse group—retirees, teachers, students of assorted ages, people with time on their hands—considered whether the site was urban, suburban, rural; whether it was tidal or nontidal; what the depth of the water was; and whether the stand was dense, sparse, or mixed. They were asked to look for animal tracks, for nests and for cocoons, for evidence of deer browsing, and for butterflies and moths. And they, too, found that the loosestrife often served as habitat. Caterpillars were found feeding in the stands, raccoons foraging in them, muskrats building lodges in them, carpenter bees buzzing through them, and red-winged blackbirds nesting in them.

One of the difficulties in evaluating a stand of purple loosestrife, he came to believe, is not what we see but what we expect to see. We come with preconceptions. "People have read and heard, over and over again, that there is no life in purple loosestrife stands. Why look for it if you don't think you are going to find it? People think it is a biological desert." Kiviat puts a high value on what the untrained eye may fall upon, on what can be seen when knowledge and expectation are replaced by curiosity and when theory is replaced by happenstance. His argument extends to a call for a kind of nonselective observation. People get into bird watching, he says, or butterflies or dragonflies or wildflowers. "And there is an exclusivity to all of this. Bog turtles are really hard to find. I really have to focus. But if you compartmentalize your perception too much, you miss the context." What he is pointing out, of course, is the paradox of what it takes to be a good witness, some alchemy of curiosity, observation, memory, diminished expectation, and knowledge, but only insofar as it is that kind of knowledge that can tolerate correction.

It may be one of the stronger arguments for folding volunteer observations into research: when participants come to a project without prior knowledge of the probabilities, they might see anything; in science as in life, when people do not know what they are looking for, they may be more open to the serendipitous find. Perhaps the most celebrated of such discoveries was made by Jilene and Jonathan Penhale, ages eleven and ten, near Arlington, Virginia, in October 2006. The kids were taking part in the Lost Ladybug Project, a citizen science program aimed at children between the ages of five and eleven, when they found a nine-spotted ladybug, the first of its kind to be documented in the eastern United States in fourteen years.

Increasingly, studies of purple loosestrife have become studies in the uncertainty of disruption. Kiviat found that red-winged blackbirds, American goldfinches, marsh wrens, and the occasional common grackle nest in the loosestrife stands. Meadow voles feed on their roots, turtles may bask on dead, flattened stalks, and frogs may use the shade the plant provides. Many insects graze the leaves, and the flowers are visited by flies, bees, moths, and almost any species of butterfly. All in all, Kiviat has documented more than two hundred insects and forty species of birds, along with mammals, amphibian, and spiders, that have used the loosestrife stands as habitats.

Jennifer Forman Orth found a similar diversity in a citizen science program on loosestrife conducted in 2005 through 2007 at the University of Massachusetts in Amherst. She invited the volunteers in her program to post their photos of any plant or animal found using the purple loosestrife on Flickr, the web-based photo sharing site. The subsequent 195 photos documented some ninety-five species of spiders, plants, snails, and insects; they revealed which parts of the plant were used—flower, leaf, stem, or entire plant; and they recorded feeding, mating, predation. All of which allowed Forman Orth to conclude that the plant is used as a habitat, that its flowers are visited by flies, bees, and butterflies, and that it provides a mating and egg-laying habitat, a hunting ground, and a resting place. Although such projects may sacrifice depth for breadth, she notes, they nonetheless can cover a wide area and engage the community, while leaving creatures unharmed.

The early September afternoon Kiviat took me out to a stand near his field office, we found the loosestrife interspersed with cattails, goldenrod, phragmites, assorted native grasses, red clover and knapweed, and white vervain. A small eastern cottonwood sapling was doing its best to assert a bit of botanical authority. The loosestrife had been mowed here over the summer, and tiny holes in its leaves attested to the presence of beetles, though it was not entirely clear what kind of beetle. Still, in a period of twenty minutes, Kiviat pointed out a tiny spider industriously constructing its funnel web, several large carpenter bees, a Japanese beetle nibbling on a flower, and four large, brown mantids, which can thrive on the insects they find in loosestrife stands. Creamy white cabbage butterflies flittered among the stalks, as did a silver-spotted skipper. A grasshopper was resting on a leaf. A jumping spider foraged in the weeds, and what I took to be a tiny, glossy dot was, in fact, a flower fly. A monarch butterfly hovered at a nearby stalk. The aroma of wild mint infused the stand. Where the soil had more moisture, the raised root crowns of the plants could have four or five species of moss growing on them.

"After a century, plants develop complex relationships with what is around them, and a vast number of introduced species never do harm," Kiviat said. "Many, like chicory, are completely benign. Less than 1 percent of nonnative plants are destructive. The problem is that you don't know ahead of time which those are." All of which helps to explain why he questions the manner in which military terms and metaphors—"invasive," "interventions," "assault"—have been adopted into the ecological lexicon. The use of such language, he suggests, reaffirms the preconception that the plant is an aggressor that needs to be eliminated. He prefers such terms as "native" and "nonnative." "Invasive can be native or nonnative," he told me. "It just means an overabundance." He suggested as well that the wetlands in which the loosestrife takes hold may already have been compromised by salt run-off, siltation, drainage, or other forms of ecological neglect and mistreatment that are a result of the stress a growing human population puts on the natural world. Which is to say, the proliferation of nonnative species may be a symptom of the problem rather than the problem itself.

The management of these species and their accelerating rate of travels around the earth are of ever increasing concern in the global age. While few conservation biologists have blanket contempt for alien species, debate persists whether to accommodate or eradicate. The argument for the former is often presented as a kind of real-world pragmatism and appreciation for biodiversity. It is a view that repositions the invasive plant as the resourceful intruder, at once resilient, tough, adaptable, and able to find a place for itself in areas that have been disrupted by human use, presence, management, and mismanagement. Human interference has *already* disturbed native ecologies. The logical extension of this argument is that the study of such robust and adaptable nonnative plants should be centered on their positive use rather than on strategies for their removal and elimination.

Even the phrase "alien species" may reflect cultural anxieties, not to mention xenophobic prejudices caused by globalization, immigration, assimilation that we have transferred to the plant world, or so the argument goes; and what we imagine to be ecological integrity is not far removed from cultural bigotry. In her 2001 paper,

"The Aliens Have Landed! Reflections on the Rhetoric of Biological Invasions," Banu Subramaniam suggested that the language surrounding nonnative species may reflect entrenched anxieties about globalization. "The battle against alien and exotic plants is a symptom of a campaign that misplaces and displaces anxieties about economic, social, political, and cultural changes onto outsiders and foreigners," she writes, and catalogs how our perceptions of and language around exotic species parallel those we apply to immigrant populations: the numbers of the foreigners, the subversive stealth with which they arrive on foreign soil, the difficulty in their control, their aggressiveness, their persistence, and their extravagant fertility patterns.[2]

Dialogue around the fear of exotic species has gone so far as to recall efforts by the Nationalist Socialist Party in the middle of the last century to eliminate foreign plants from German soil. Hitler's advocacy of ecological engineering, for example, famously banished *Impatiens parviflora,* viewed then as a kind of botanical Bolshevik, "a Mongolian invader" capable of threatening Germany's entire Occidental culture. Notwithstanding questions about how so frail a shrub could carry such a burden of botanical nationalism, the ideology of blood and soil suggested that how plants take root in the landscape expressed basic truths about national identity.

Metaphor surely serves a purpose as a communication device. As Brendon Larson writes in *Metaphors for Environmental Sustainability,* "Metaphor is a key element in scientific inquiry because it enables us not only to understand one thing in terms of another but also to think of an abstraction in terms of something more concrete and everyday. . . . Metaphors also help us interpret the novel and the unknown by invoking our shared cultural context." But that said, metaphors are limited in their use and clarity; they are open to interpretation, and their implicit analogies can be imprecise and partial. Like any other living thing, they have a limited life span.[3]

It is hard to attribute caution toward nonnative species to xenophobic hysteria, a bit far-fetched to assume bigotry on the part of biologists concerned about local ecologies. There is sound science for the case that some nonnative species visit damage upon their new host habitats: water chestnuts *do* deplete oxygen from areas of the Hudson River, the woolly adelgid *has* decimated hemlock trees, and the Asian longhorned beetle *is* a genuine threat to ash, birch, and maple trees. The problem with nonnative species is that no one knows what they will become or what their effect will be, which will be benign and which more likely to injure host habitats. It is in the nature of the new to contain mystery.

But even apart from this, assumptions about our xenophobia seem questionable. Most of us, after all, have a complex relationship with the unknown. If it makes us fearful and apprehensive, so, too, does it engage us. Some of these nonnative species were imported to this country exactly because we find such allure in what is foreign to us. There are times we operate from the conviction that deviating from the ordinary and familiar will deliver us to wonder. Consider the American Acclimatization Society, whose well-intentioned efforts to cultivate our literary imagination, if not ornithological acumen, drove it to try to introduce every species of bird mentioned by Shakespeare to American soil and skies in the late nineteenth century. In 1890, one of its members, Eugene Schieffelin, released one hundred or so European starlings—noted by the Bard in *Henry IV*—in Central Park. What we have to show for it today, of course, are some two hundred million noisy, intrusive nuisance birds, but who then could have argued with the notion that such a gesture might fill the air with a dazzling, new chorus of birdsong?

If we have a prejudice toward nonnative species, surely it is balanced by our willing embrace of the unfamiliar; we are just as capable of being amazed and beguiled by the exotic. It is why I drive into my driveway and find four peacocks sitting on the roof of my house or why my neighbors raise emus and llamas. The allure we find in plumes of giant bamboo or sprays of Chinese silver grass or any number of other nonnative flora in our own backyards and gardens is based on our human attraction to what is new to us. It is in our contrary nature to be drawn to what does not belong to us, and when it comes to the natural world, our sense of partisanship is inconsistent, unpredictable, and erratic. At once enchanted and threatened by the foreign, we humans revel in ambivalence and inconsistency when it comes to the unfamiliar.

On the August afternoon I visited the stand of loosestrife down at our own corner, I simply heard the steady drone of bees and the hum of crickets and saw a half-dozen eastern tiger swallowtail butterflies. The loosestrife here was in a mixed stand, part of a late summer collage of cattails and willow, a few maple saplings, ferns, jewelweed, wild carrot, and goldenrod. While conventional wisdom suggested this wetland was inert, it was clearly thrumming with energy and life. A couple of mornings later, two white cabbage moths flitted from stalk to stalk. It was not hard to see how this little patch of wetland could be as much about placement as it was about displacement.

A quarter of a mile down the road was another stand of dense loosestrife that looked to be monotypic, though a closer look revealed an undercover of bog grass clumps interspersed with delicate ferns. Perhaps the beetles that seem to have thinned the first stand, allowing it to integrate with other late summer plants, had not yet arrived here. The protocols of biological controls are elusive, and infinitesimal differences in soil, moisture, drainage, and light may affect their work. What takes two summers to succeed in one place may take twelve summers just down the road. My neighbor whose house sits just behind the stand doesn't know what the plant is called, nor does she have any idea of the horticultural contempt it so often meets. "It's pretty, isn't it," she asked the morning I stop by. "Sometimes I see a heron in there." A year from now it may look entirely different. "There is no end of the story in ecology," Kiviat has said. "There are only probable outcomes."

Nothing about purple loosestrife designates it a representative of exotic species. Each nonnative plant, bird, or mammal that arrives on a new woodland, field, or riverbed reacts with its own characteristics to its own particular circumstance. But the myriad conflicting assumptions and reactions toward it can speak to the ambivalence with which we so often fold the unfamiliar into our lives. Given that the global ecology is changing rapidly, it may be time to acknowledge the complexity in our own attitudes and reactions. How we tend to the places we care about may depend on acknowledging such intricacies. A place to start may be with the term "neobiota" for introduced species. Dry jargon though it may be, it is unlikely to inflame questions of cultural bigotry and probably has a place in a time of urgency, of climate change, habitat loss, and patterns of extreme weather, all of which understandably elicit strong reactions and strong words.

But more important than such measured language is our measured response to these species. David Strayer, a freshwater ecologist at the Cary Institute of Ecosystem Studies, suggests that we address nonnative species with a consideration for the timetable of their arrivals. For those plants that are already here, he suggests local management strategies: specific programs with specific goals for specific plants. And for those plants just arriving now, he advises control whenever possible, whether they are calculated biological controls or more intensive eradication initiatives. Finally, for those species that have not yet arrived, it is his hope that technical, legislative, and educational strategies will prevent them from ever getting here.[4]

It seems a sensible plan of action. Watching the butterflies threading their way through the purple stalks on a September afternoon, I know this is a single scenario of many. By mid-September, the loosestrife will fade to brittle russet stalks, its flowers becoming dry, its vibrant magenta turning to rust. With its spectacle of color subsided, its new autumn palette will be absorbed by everything that surrounds it, and it is tempting to imagine what is in front of me a landscape of management and adaptation. But as new species continue to arrive at an unprecedented rate, some of them may pose a more rigorous challenge. If nothing else, then, the bank of purple loosestrife might simply serve as a reminder about our ambivalence toward the exotic; how it is possible to be both engaged by and fearful of what is strange to us; how awe and anxiety can be such natural colleagues; and how it is fully human to both fear and admire what is not familiar to us. And I think of Mrs. Bacon and her Russian bobcat and all those ways we awaken to the unknown when we think we spot it, streaking across the meadow at dawn.

9
Eels in
the Stream

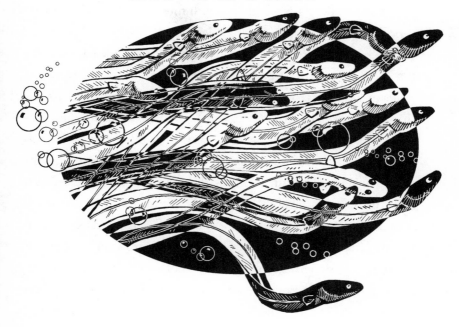

Kids love catching things, and if what they're trying to catch is in the water, that makes it even better. It's better yet if you happen to love what you are trying to catch, which is where Chris Bowser enters the picture.

He's the first to admit that he's in love, and to prove it, he passes around the objects of his affection, four slender glass eels, each only two or three inches long, swimming in a small glass jar of water. Which is to say, you're looking at something that looks like nothing swimming in something else that looks like nothing. But Bowser is radiant and doesn't hesitate to confess his feelings. "How can you blame me for being in love?" He asks. "What about you? How can you not be in love with these?" Suddenly things are very Zen.

An environmental scientist, Bowser is the science educator for the New York State Department of Environmental Conservation Hudson River Research Reserve and Estuary Program, in partnership with Cornell University's New York State Water Resource Institute. Today he is addressing a group of students from Poughkeepsie High School who have volunteered to help him with the annual spring eel count. Along with high school science teachers and students from nearby Marist College, the kids converge every Monday afternoon for two months at the Fall Kill Creek, a tributary to the Hudson River in Poughkeepsie, to monitor the eels swimming upstream. The program has both conservation and education goals. While Bowser is teaching the kids about the transoceanic journeys of the American eel, water quality, ocean currents, and the food chain, he is also doing his best to pass along his own personal fondness for the creature with efforts that extend to the culinary; standing in the creek in his waders, he's not beyond slipping a scud, an infinitesimal shrimplike crustacean, into his mouth, cocking his head, seeming to savor it as any astute gastronome might, then announcing the flavor as "mud-seasoned shrimp," before he launches into further explanation of the eels' life cycle.

American eels are born in that area of the Atlantic Ocean known as the Sargasso Sea, but scientists have never been able to document their mating or birth. Born in salt water, the eels then migrate to freshwater rivers and streams, returning to the sea years later only to spawn and die, in a cycle called catadromy. Their lives begin as tiny planktonic larvae in the shape of willow leaves. As they drift toward the coast, borne by currents into the freshwater rivers and tributaries on the East Coast from Florida to Greenland where they will spend their lives, they are transformed into slender, transparent "glass" eels, only a few inches long. Traveling up the tributaries, they continue to grow, and their pigmentation turns a darker greenish brown, and they become "elvers." Larger maturing eels, or "yellow" eels, live in freshwater tributaries; female eels, which can grow up to three or four feet long, can live in the tributaries for up to thirty years. Only toward the end of their lives do they become "silver eels." Then, their eyes grow larger, possibly to enable them to discern new species of predators, while their digestive tracts shrink to make room for eggs, or milt. Their skin turns from a greenish color to gray, cream, or silver, and their bodies thicken, all in preparation for the rigors of the return voyage to the sea.

In recent years, though, the eel population has declined, possibly because of contaminants in the water, dams, overfishing, parasites, habitat loss, or hydropower projects. Possibly, too, changes in the ocean current caused by a warming climate are the cause. Or perhaps it is something else entirely. It is a question the Atlantic States Marines Fisheries Commission would very much like to answer. It's also a question that has spawned some smaller ones: What environmental factors affect eel migration? Water temperature? Tide cycles? Precipitation? What affects the eel population site to site, year to year? Do they migrate steadily or all at once? How many of them are there, and how are these populations changing?

It is to help him answer such questions that each spring since 2008 Bowser has recruited a corps of volunteers, among them students, teachers, interns, retirees, and watershed citizen groups, to count the eels entering ten Hudson River tributaries from Yonkers to Albany. "One of the great things about studying eels," he says, "is that by paying attention to them, we are paying attention to the entire ecosystem. Eels rely on oceans, estuaries, streams, and entire watersheds. They are found in farm ponds and city creeks. They are the perfect ambassador for the entire ecosystem because they *live* the interdependence that environmental scientists like to point out. This is especially important to the Estuary Program and Research Reserve, which seek to link all parts of the system together—from scientific, management, and conservations perspectives."

To trap the glass eels as they enter the tributaries, Bowser and his colleague Sarah Mount install a fyke net early each spring at the mouth of each of these tributaries. Twelve feet long and funnel-shaped, the net is designed as a series of connecting hoops diminishing in size so that it will catch only the smallest waterborne species as they travel upstream. The net is staked to the streambed with rebar and a concrete block to keep it in place for the two-plus months of migration. Once a day, volunteers at each tributary collect what has gathered in it, counting, weighing, then finally releasing the tiny eels further upstream, past the dams and barriers that might hinder their migration. And every eel released has a better chance to make it back to the Sargasso Sea to spawn. Since 2009, more than eight hundred volunteers have released approximately a hundred thousand juvenile eels above dams and other barriers to migration.

To reach the net in the Fall Kill isn't just a question of stepping through slippery stones in frigid water, a process Bowser describes more or less accurately as "just kind of weird, like you're shrink-wrapped and walking on greasy cannonballs." It can also require ne-gotiating your way around plastic bags, candy wrappers, soda cans, Styrofoam cups. Anyone stepping in the water to check the net might find themselves stepping around an old bicycle tire or shop-ping cart. Which is to say, the Fall Kill is an urban tributary. While it begins in Hyde Park, flows through the towns of Clinton and Pleasant Valley, passing through woods, wetlands, marshes, grass-lands, and residential areas, by the time it flows into the Hudson River sixteen miles later in Poughkeepsie, it's become a city water-way. On its northern bank are the crumbling remains of a stone wall constructed during the New Deal to channel the water through the city; above the wall now is a chain-link fence that teeters at angles that could only have been determined by seasons of wind and rain. Whatever biological life is in the water running directly beneath it must coexist with all the other detritus that blows, rolls, tumbles into the water or is put there because people can't find any better place for it.

But if an outing to the Fall Kill doesn't readily conform to our ideas of how nature is best experienced, what's happening here with the Eel Project reflects the way conservation is likely to be practiced as the world's population continues to move from rural to urban areas; since 2008, 3.3 billion people, or half the people on earth, have lived in cities, a population shift that is only likely to con-tinue. Although recent generations of environmentalists have tended to focus their efforts and enthusiasm on areas of rural wilderness, future ones will necessarily address the ecological renewal of cities, whether through planting trees, restoring urban waterways, or simply being attentive to the animal species that seem to show up with such, albeit unexpected, regularity. As urban ecologist Steward Pickett has pointed out, "Processes we associate with natural areas go on in cities. Plants grow, organic matter decomposes, and water transports nutrients and contaminants. Streams harbor invertebrates and fish, air carries particles and gasses, some of which affect human health, and animal diseases are spread to humans."[1]

To remain an effective enterprise in the coming decades, conservation practices may increasingly depend on how city dwellers can come to recognize the nature of their own habitats. The kids at the Fall Kill today don't seem to have much trouble with that. For many of them, stepping into the stream *is* a rare excursion to the natural world. It's not some synthetic environment that has been shaped, designed, digitized, or Disneyfied, not a touch screen, virtual experience, or augmented reality, but a river flowing one way, with eels coming the other way, and a net positioned at their convergence. What happens beyond that is going to be anyone's guess. As Mark Angevine, a Poughkeepsie High School teacher who partners with Bowser at the Fall Kill puts it, "A lot of these kids have only experienced nature by watching TV. Or in controlled places, like the beach, where there are lifeguards, or areas of water that have been roped off. And they've never been down to this brook before." Angevine has been working with the program since its inception. Notwithstanding the stud in his ear and an inclination toward checkered shoelaces and graphic ties, Angevine has a wry humor and matter-of-fact approach that make for a supportive presence for the kids who may not be altogether sure why they are putting on waders and heading into frigid water.

And why they would do that exactly doesn't have simple answers. Possibly it's an interest in biology, or the extra credit, or the chance to hang out with friends outdoors. Kelly Lattin, a living environments teacher from the high school and another Monday regular, suggests that it's about "getting them to notice what's going on around them. I don't know if they take nature for granted, or if they don't even know it's there. . . . This is just huge exposure for them." And possibly, too, it's some basic curiosity about why things are the way they are. Or maybe it's just that teenagers, by nature and definition, are clued in to the fact that a lot of things in life seem to generate more questions than answers. Hold a glass eel in your hand, and it slips through your fingers like water, its slight physical being every bit as elusive as all the other facts of its existence. When Bowser puts one into a student's hand, Jen, who is fifteen, steps back reflexively, then laughs as the little sliver of mystery squiggles around in her palm. The Eel Project, in other words, is perfect for them.

If science is about finding answers to those things that are not known, Bowser's interests seem clearly weighted toward the questions rather than the answers. Surprise and speculation are at the core of his curriculum, and he prefaces more than one conversation by saying, "To the best of our scientific knowledge, and there are a whole lot of gaps in it . . . ." He traffics in conjecture and revels in uncertainty, and it is exactly this full engagement with the unknown that makes him perfectly matched to his subject, which is the pure mystery in the life of the American eel. And if he begins with a mode of scientific speculation, when he is with the kids, it escalates toward rapturous incredulity; when he quizzes them about where the glass eels come from, one small voice ventures, "The Atlantic Ocean." He nods his head, but it is in dramatic disbelief. "This little guy swam twelve hundred miles to get here. That's *unbelievable. Can you *imagine* that?" My own guess is that this is a view to things that is especially suited to teenagers, whose outlook on life tends to veer wildly and unpredictably between unshakable conviction and untroubled oblivion.

Once the kids have put on the chest waders and are out in the creek, they check the net, which involves detaching the ropes and carabiners that secure the net to the rebar and cinder block anchor, unknotting the ropes at the tarp end, looking into the net, and taking out whatever happens to have collected there. If eels have turned up, they're put into a bucket so they can be taken to the shore, counted, and weighed, weight being an indicator of their health. The eel mops need to be checked as well. Like its colleague in the kitchen, the mop is simply a weighted plastic dish with a head of tough synthetic strings attached to it. Immersed in the water and secured with bungee cords, it can be left in place for weeks at a time. When the eels find their way to its welter of strings, they may stay, mistaking them for the obscure tangle of dark weeds where they might otherwise take up residence. Checking the mop requires putting the mop head in a large tub of river water, plunging it and swishing it around to get the eels to release themselves from the strings they've fastened themselves to, then pouring and filtering the water from the tub into a net that will catch the eels so that they can be counted. It's an alternative to the net that's especially useful in waters too deep or too fast to wade through; you can't have too many ways to get information about eels.

But on this particular early spring day, the counts from both the net and the mop remained low. Six glass eels and four elvers. The winter of 2011 had been especially harsh; the temperature had been warmer last year. Bowser conjectured that *maybe* it was about the weather. And the salt front had remained well to the south. *Possibly* it was easier for eels to get the push from the salt front as they move upstream. But that's just a *guess,* he said. And at high tide next week when the moon is full, it *could* be that the number of eels will soar. But who knows? One of the reasons, possibly even the main reason, kids pay such close attention to Bowser is this keen appreciation, profound respect, possibly even exuberant reverence for the inexplicable. And even if the count is low, the inscrutability of eels remains a source of endless fascination. "Eels have such a compelling story," he said before he sends the kids home. "They're born in the ocean but spend their lives in the watershed. That is so weird! Eels are so charismatic . . . in a Humphrey Bogart kind of way. They don't look so good, but they are really fascinating."

Bowser spent two years in the Peace Corps in Mauritania, where most of his time was spent planting trees in the Sahara Desert. "For the Mauritanians I worked with," he explains, "this was the real deal. It wasn't some abstract way to do good. This was a desperate act to combat the disappearance of resources. Planting trees was absolutely essential." As skilled as he was in motivating other people, he found that he was himself equally motivated by the exchange. Eventually, back in the United States, he ended up as the education director of the *Clearwater,* the fabled sloop that travels up and down the Hudson River promoting environmental education, stewardship, and advocacy. He subsequently went to the Department of Environmental Conservation, where he worked as an educator, researcher, and mapper, initiating the eel program in 2008, based on methods previously established by eel researchers in the Hudson Valley. "The Marine Fisheries Commission wanted this information," he says. "And the Hudson River Estuary Program and Research Reserve decided to fund it. But the really, *really* innovative part of this program came when we realized we could do it with student volunteers." Critical support followed from more than thirty local organizations, including nonprofits, schools, and watershed groups, along with grants from the National Audubon Society and Toyota's "Together Green" program.

Bowser points to three distinct value points of the Eel Project. The first has to do with the count—how many eels are there, where they are, and when they're there. That's what Marine Fisheries wants to know. The second thing is which site gives you the most information, which tributaries will benefit most from restoration efforts. Or, where can we do the most good. And third, for volunteers, is the service learning component. "The big value of citizen science is that it connects people to the environment and gives them confidence that they can do something about it."

Throughout April, the temperature remained cool and the rain steady, and the Fall Kill continued to run fast and high. Is climate change a factor in the diminishing numbers? What is the effect of climate change on eels? "This is a very rugged animal. They've been around for eighty or ninety million years," Bowser told the kids. "So there are a lot of questions. People say that with climate change, it's going to get wetter and we're going to have more intense storms. I dunno. But it's interesting to check eel counts after this kind of condition." Bowser is reluctant to take climate change head-on, finding it "more useful to think of the smaller parts. That makes it personal for people. It supplies more useful information."

And consistently, the information was in the low numbers. On the third Monday out, the kids found nine glass eels and one elver. "Imagine being an eel," said Bowser, looking at the torrential stream. "Imagine trying to fight this thing now. My guess, and it's only a guess, is that the water is finally going to get warm. There's a big full moon coming. And it if doesn't rain, in forty-eight hours, there's going to be a big jump in eel numbers. Maybe even tripled by next Monday. Eels are in the Hudson, just waiting to come in!" Then he asked the students to guess at the numbers for next week: 38, 39, 45, 25, 19, 12. "I'm guessing 72," he said. The eventual daily high for the season was 52 glass eels, low for the Fall Kill compared to recent years, and far below the one-day record of over 500.

But the next week, things weren't much better. The rain hadn't subsided much, it had remained cool, and the Fall Kill was still running with turbulence. Exactly three glass eels showed up in the two mops and zero glass eels in the net. Martice, at seventeen, already a man of few words, simply said, "Maybe it'll be better next week." Daniel, who was fifteen, added that "it just shows you how things always change." Already both of them speak with the acceptance, succinct philosophizing, and terse realism that seem mandated in this line of work. They seemed to understand that the vast enigma of the American eel is best addressed with straightforward pragmatism.

The natural affinity they have for their work at the Fall Kill evoked their distant predecessors from another century in another state. In *The Log from the Sea of Cortez*, John Steinbeck documented an expedition to gather specimens of marine invertebrates in the Gulf of California. The team, an unlikely gathering of biologist Ed Ricketts, Steinbeck, and assorted crew members of questionable background and motive, had been gathering sea cucumbers, brittle-stars, starfish, anemones, sea urchins, crabs, shrimps, worms, limpets, snails, and sponges, but on their arrival in La Paz, Mexico, they discovered a legion of untrained but unexpectedly lively and competent assistants suddenly available to them. Steinbeck observed:

> As always when one is collecting, we were soon joined by a number of small boys. The very posture of search, the slow movement with the head down, seems to draw people. "What did you lose?" they ask.
> "Nothing."
> "Then what do you search for?"

Small boys, he continued,

> are the best collectors in the world. Soon they worked out a technique for catching the shrimps with only an occasionally pinched finger, and then the ten-centavo pieces began running out, and an increasing cloud of little boys brought us specimens. Small boys have such sharp eyes, and they are quick to notice deviation. Once they know you are generally curious, they bring amazing things. Perhaps we only practice an extension of their urge. It is easy to remember when we were small and lay on our stomachs beside a tide pool and our eyes and minds went so deeply into it that size and identity were lost, and the creeping hermit crab was our size and the tiny octopus a monster. Then the

waving algae covered us and we hid under a rock at the bottom and leaped out at fish. It is very possible that we, and even those who probe space with equations, simply extend this wonder.[2]

As the narrative continued, the boys came out in canoes and flat-boats, others swimming, all of them energetic, able, inquisitive, and above all skilled at finding and delivering the aquatic specimens. Natural hunters in the habit of looking for unknown things, they thrill to those moments when their innate curiosity can be put to a practical purpose.

The Fall Kill isn't the tranquil turquoise of California coastal seas but a chilly, gray stream on a raw spring day in the Hudson Valley, and there were as many girls here as boys, but the kids from Poughkeepsie High exhibited a like initiative and enterprise, and it seemed likely that their willingness to put on chest waders and step into the frigid waters was fueled by a like sense of wonder. Only yards downstream, the Fall Kill drained into the Hudson, and that wide, shining corridor streaming along our far field of vision offered a frame for the entire enterprise.

Besides, a sense of camaraderie was building down at the river, and even if the eel count wasn't adding up to much, something else was. It's a human tradition to build at the river: mills, factories, houses, towns, cities. Under construction here was a different manner of community. As Angevine points out, the project serves a population diverse in both ethnicity and academics. "You have AP biology students here who have a real interest in science working alongside kids who may have learning disabilities or autism. Out here, they're all working with each other, along with college students, and people from the Department of Environmental Conservation. Out here, they're all part of the same team."

Almost on cue, Jorge, a student at Dutchess Community College, showed up to help out. Twenty now, he had started with the first year of the eel program as a junior in high school. Now he just comes year after year, whenever he can. "Just being here," he said, "seeing how many eels there are, we're in New York, how many travel from so far away . . . ." He waved his hand at the wide river and his voice trailed. He was a major in environmental sciences with a second in geology. That he had been at it from the start was clear; he moved in the water with easy confidence only partly due to his footwear—neoprene and rubber-webbed water shoes. His ease and balance on the slippery rocks came from something else as well, maybe nothing more than a certainty that he was in the right place. When the high school kids needed help getting the eels out of the mops, he was right there, and when he'd done with that he didn't hesitate to take the scrub brush and start cleaning the fyke nets.

But it wasn't just the kids, their teachers, and Bowser at the Fall Kill. Maybe waterfronts invite people who are curious, or maybe being at a waterfront just brings out some innate human impulse to be curious. Who knows exactly how human inquisitiveness and wonder arrive at their intersection with the riverfront, but the anglers at the Fall Kill took a regular interest in what the kids were up to, and one afternoon, a city engineer swung by to give the kids a primer on storm drains, illicit discharges, water turbidity, and how to protect waterways; afterward he hung around to see what turned up in the fyke net. Curious onlookers drifted over as well. On another afternoon, a guy in plaid pajamas, sandals, and a bright blue

T-shirt that read, "If you met my family you'd understand," wandered by and found himself immeasurably pleased by the combination of tenacity and serendipity that have delivered the eels to this particular destination. He shook his head and marveled in delighted disbelief to no one in particular, "Over a thousand miles in the ocean current to get, of all places, to *Poughkeepsie*."

But none of this changed anything. The three-digit count everyone had been hoping for remained elusive. The following Monday, the fyke offered up 5 elvers and 14 glass eels. And even by the end of May, the counts didn't come up much. By the end of the season, volunteers had counted 625 glass eels and 218 elvers at the Fall Kill. Whereas a year earlier, the fyke net might have produced over 100 eels on any given day, the last Monday this team was out, the net produced exactly 3 glass eels and 3 elvers. And when Lattin's thirteen-year-old daughter, Tyra, came in from checking the mop and gravely announced that she had 13 glass eels in the bucket, she was met with incredulity. Bowser could scarcely believe the discrepancy between the count in the net and the mop, and he lapsed into the technical parlance that so often seemed to attend such discoveries. "This is soooo weird," he said, and it seemed only right that this last day of the eel count ended with yet another question.

Eel counts had been slightly higher upriver, possibly because the cool, rainy spring and low water temperatures had encouraged the eels to stay in the main stem of the river longer; when they finally entered the tributary, it was at a more northern location. But the possible reasons for the overall low count vary. The rainfall from March through May this year had measured in at fifteen inches, eight inches more than the previous year, and that alone could have been the reason for the eels' trouble in reaching the tributaries. Or possibly excessive snowfall during the winter had kept the streams unseasonably cool late into the spring. Or maybe some unknown environmental factor in the Sargasso Sea had delayed the eels' spawning. Or perhaps it was something else. The kids knew by now that getting high numbers is not what the Eel Project is about; they'd gotten the core lesson: whether it's three glass eels, thirty, or three hundred, it's all data.

The behavior of eels has such a grand inscrutability that expectation is replaced by curiosity and an abiding respect for fact: what they do, when they do it, where they do it. Jen summed it up when she said, "Each week you expect more, but it's never there. So you just gotta keep hopin'." Then added: "I'm definitely gonna do this next year. First, 'cause it's *fun*. And it's extra credit. And it's something I'm doing for my community. And then I'm just gonna learn more about eels than you ever knew was there." Something about the idea that there is so much yet to learn about this tiny ocean creature made her burst out laughing.

And I realized then that while the kids are learning about glass eels, water quality, the food chain, and river ecology, what they were also learning is about that possibility of folding uncertainty into their lives. Searching out the unknowable is likely to be one of those things that define us as human beings; and how we accommodate the inexplicable into how we live will do much to determine the quality of our adult lives. James Prosek considers this in his book about eels, writing, "We allow ourselves to believe that nature can be explained. In the process, we confine nature to those explanations. The eels, through their simplicity of form, their preference for darkness, and their grace of movement in the opposite direction of every other fish, have helped me to see things for which there is no easy classification, things that can't be quantified or solved, and get to the essence of experience."[3]

Now, on a spring afternoon at the Fall Kill, that there were more questions than answers suddenly seemed like the greatest lesson of all. No wonder Jen was laughing. But it was the end of the day now, and the kids were getting ready to head home. Bowser thanked them, but with an inclination toward effusiveness, he didn't just thank them once in English. *Adios! Au revoir! Sayonara! Auf Wiedersehen! Arrivederci! Ma'salaam!*

In their study of cultural differences in how we perceive our place in nature, environmental psychologists often consider the figure-ground relationship. In the history of Western art, the human figure and the natural world are often represented as separate entities; the figure is used as a measure of the scale or power of the natural landscape, as a counterpoint to it, as an observer, as a separate presence and force. In Japanese and Chinese art, however, the separation is less distinct; the monk, the cabin in the woods, the fisherman in the boat—all of these are integrated into the natural world. One textbook I have notes that "in Chinese art, the problem of depth was handled by using space to imply more space beyond the boundaries of the painting. The scenes shade off imperceptibly in mist-clouded rock masses, in forms half-revealed. Distance is suggested rather than calculated. The finite structure common to the Western view of nature gives way to a world that is unlimited and ultimately unknowable. The spectator is not outside looking at the painting; he is part of it."[4] Now, watching the kids counting glass eels on this slightly overcast spring afternoon, it is easy to imagine a like coalescence of figure and ground. "In Eastern landscape painting the activity, the style, scenery and people are often seen as one mode;

strictly speaking, there is no figure and ground," my textbook notes.[5] Certainly that convergence is what I seem to be witnessing today.

Bowser may be uncertain about how eels get here, and he may not know just how salt fronts or water temperature or tide figure in the voyage. And whatever assumptions he *does* have are challenged regularly. "We used to expect to see high migration numbers at the new and full moon," he says. "But we didn't see that." He sighs. But one thing he is certain about is that the Eel Project is a good model for citizen science, first, because this is a species with a genuine conservation need. Second, from the volunteer point of view, it's a contained project: fyke nets are checked every day at a fixed time for two months. Third, it has strong links to management agencies, and finally, there are multiple data points, meaning numbers of eels, their weights, and water temperature; anyone working on the project is likely going to see something. Projects are most rewarding, especially to kids, when there are results.

But I would suspect that the project's success has to do with mystery as well and its comprehensive lesson plan for ambiguity. "When a scientist doesn't know the answer to a problem, he is ignorant," the Nobel-winning physicist Richard Feynman once stated. "When he has a hunch as to what the result is, he is uncertain. And when he is pretty darned sure of what the result is going to be, he is in some doubt. We have found it of paramount importance that in order to progress we must recognize the ignorance and leave room for doubt. Scientific knowledge is a body of statements of varying degrees of certainty—some most unsure, some nearly sure, none *absolutely* certain."[6]

"Look at that delicious little thing," Bowser had said to the kids early in the season, holding a slender glass eel in his hand. "If you look closely, you can see what it's eaten lately, its tiny eyeballs, its little heart."

How is it possible to hold something so utterly small and transparent in the palm of your hand and still know so little about it? We like to think of transparency as a means to knowledge; when something has clarity, it is revealed. Unless it's an eel, that is.

Herring are more likely to reveal themselves in water that is shadowy and obscure, I have learned, while the glass eels, completely translucent, tell us impossibly little of themselves. In this tutorial on observation, this sidebar on clarity seems important. Maybe even significant. And I wonder if this is what most of all makes this project so engage the human imagination. Maybe any entity with so obscure a beginning and so indefinite an end, with so mysterious an instinct to travel so far, guided by such uncertain influences and impulses, is bound to speak to us. Maybe there is something in the life cycle of the eel that speaks to the way we ourselves work to navigate the wide waters between the unsure, the nearly sure, and the none absolutely certain; and to the unknown we all feel about where we start and where we end, about our origins and our final destination and what forces deliver us from one to the other. If kids like catching things, and catching things in water even more, something in the persistent enigma of this particular catch makes it irresistible.

Postscript: The following season, 2012, after a remarkably mild winter and dry spring, the net was back in the Fall Kill and numbers skyrocketed to 6,751 glass eels and 198 elvers. Perhaps more important, almost everyone mentioned here, from Bowser and Sarah to Jen, Martice, and Daniel, was back again, still enthralled, still cradling little eels.

10

Vines Through the Trees

Things happen too quickly. Life happens too fast. Time passes at an accelerated rate. We know this and complain about it. "Hurry sickness" is the name sociologists have given to the chronic rush that so often attends our lives, that imperative so many of us have to do the next thing now. All of which is why mile-a-minute vine could be a plant invented for our age. It, too, has hurry sickness. Named for the speed of its maniacal growth cycle—up to six inches a day and up to twenty feet a year—the vine defies the traditional cycles of time we tend to associate with the natural world. Advancing with its own promiscuous speed, mile-a-minute vine doesn't just disrupt the landscape and native ecology but dismantles our notions of time. It is evidence that even in nature things can change right in front of our eyes. Turn your back for an instant, and the landscape has been transformed.

In our neighborhood, the vine had raced its way along the line of Jackson Creek and the woodlands on either side of it. The creek is a tributary of Sprout Creek, which flows into Fishkill Creek and then into the Hudson River, but the creek isn't going much of anywhere on this September afternoon in 2010. After a summer drought, its bed is dry, and the indiscriminate pattern of its dirt and gravel scoured by the rocks, small stones, washed-out, old tree trunks and branches is all that's left to suggest the torrents of spring rain and snowmelt. Thickets of deer tongue grass suggest that the moisture is just beneath the ground, but the water itself stays out of sight for a quarter mile or so as the streambed winds its way through a shady grove of sycamore, oak, and maple.

Yet when the trees thin and the bed opens onto a sunny meadow, it's not just the arid cut in the earth that's devoid of life. On either side of it are skeletal trees, their leaves long gone, their limbs stark fingers. The trunks, though, are sheathed in dense mats of the vine that make for a kind of botanical freak show; logical order here has been reversed. The upper limbs seem to have died, while the trunks beneath them are matted with thick, emerald foliage.

Such are the irregularities of mile-a-minute vine. Also known as Asiatic or devil's tear-thumb, *Persicaria perfoliata* has thin leaves shaped like equilateral triangles and delicate stems covered with tiny barbs. The annual vine uses these barbs to attach itself to, climb over, and blanket whatever lies in its path, wholly transforming the shape and identity of the familiar like some deranged topiary. The vine's stem is segmented, and at the start of each new segment, it can change the direction of its growth, which sometimes makes its path appear disjointed and erratic. Small, saucer-shaped leaves, called ocreae, surround the stem at regular intervals, and clusters of small berries, green at first, then blue when ripe, emerge from these nodes. And if the vine grows quickly, these berries are agents of its equally swift spreading—they can be washed downstream to generate new outgrowths or be scattered by the birds that eat them.

As the creek bed meanders farther into the meadow, the vine has woven itself through multiflora rose bushes and wild grape, lambsquarters and wild nettle. Mile-a-minute vine loves the sun, and though it can be shaded out by other plants, once it reaches the light, it thrives, spreads, and forms a canopy over other plant life. Its cloud of leaves prevents photosynthesis of those below, then smothers and eventually kills weeds, shrubs, and entire trees, transforming wetlands, roadsides, forest edges, and open fields. Here it has wound itself around the spiky pink flowers on bushes of pale smartweed, trails through a thicket of goldenrod, winds through a patch of jewelweed. A small, spindly birch is halfway to dead, the heart-shaped leaves on its upper branches laced with the vine, and the stands of timothy and tall fescue below have been overtaken. Some of the leaves of the vine have turned a pale peach color, and intertwined with fading stalks of purple loosestrife, they create a picturesque bouquet of defeat. And though it's late summer, only a matter of weeks until the first frost, elsewhere small, fresh green leaves continue to climb over, through, across whatever lies in their path.

At first glance, there is no real telling what's going to come out ahead in this riot of weed, this botanical free-for-all. But look up and around and see the trees dying with their trunks choked, and the vine's tenacious intent is clear. In some cases, the vine is layered with a tangle of last year's vines, and those of the year before are woven in, too, as though, like so many other aggressors, on top of everything else it has to contend with, it must contest itself and all its own previous damage as well. There is something in this vine that reveals the depravity of speed.

The vine is thought to have first arrived in the United States in the 1930s, when a garden nursery in Pennsylvania imported decorative shrubs from Japan, and since then, it has spread more than three hundred miles to eleven states in the Northeast. It was first detected in La Grange, New York, when the town's baseball fields just south of the creek flooded in 2006. The fields were built on the flat acreage of the creek's floodplain, so as might be reasonably expected, they flooded during a spring of heavy rains. When a local representative from the Department of Environmental Conservation visited the site to assess the damage and make recommendations for remediation, he noticed the vine and took a sample. Speculation was that the seeds may have arrived on the tires of a construction truck delivering materials to a nearby development site. By then, the vine had already started to spread.

A year later, as a result of both the creek's continued erosion and flooding *and* the infestation of the vine, a joint team consisting of the Fishkill Creek Watershed Committee, the county environmental council, and local Conservation Advisory Committee members organized the Jackson Creek Streamwalk to assess the state of the creek and determine the extent of the vine's spread. The subsequent report itemized such factors as bank erosion, litter, excessive sedimentation, and diminished vegetation on the banks, along with a series of dams and culverts that prevent fish migration. And it concluded that poor land use decisions had allowed for excessive water runoff into the creek that eventually compromised it. Streamwalk volunteers also noted the proliferation of invasive plants, including barberry, multiflora rose, phragmites, and garlic mustard. Most damaging was the mile-a-minute, draping itself over trees, bushes, and stream embankments. In a stream area already compromised, trees and bushes are necessary to stabilize the banks; when these are strangled out by invasive weeds, the stream is further degraded.

Rick Oestrike, a volunteer member of the Fishkill Creek Watershed Committee at the time of the Streamwalk, took me down the creek bed three years later, in the summer of 2010. A geologist, he is taciturn by nature, taking the greater pleasure in the language of the streambed, and he is able to read its surface, all its hollows, its configurations of rocks, its patterns of abrasions. "Streams aren't just water," he said. "They are also rocks, gravel, and debris that cause abrasion." Changes in land use, including more paved areas, diminished woodlands, and storm water runoff that travels more quickly to the stream, have all contributed to the stream's periodic flooding. Now Oestrike pointed to a large basin that had been scooped out in the creek bed, likely carved by a swirling eddy of water and a rock a foot-and-a-half long left there by spring rains. The bed had been scoured by the stones. When we reached the dryer area of the streambed where the earth seemed bleached by the sun, it was easy to see where the vine had done its damage. Although the vine had been unable to reach the tops of tall trees in a single season, ensuring the safety of the poplars and willows, smaller saplings were more vulnerable.

Oestrike's innate reticence seemed infused with a sense of resignation at the difficulty in instituting a management strategy for the weed, and he suggested that the failure to contain it was systemic. After the 2007 Streamwalk, Bill McCabe, then a Dutchess County legislator, brought recommendations for the control of the vine to the legislature in March 2008. Defining the vine and mapping both heavy and moderate infestation areas, they suggested strategies for eradication, whether by private landowners, with town assistance, or by town order. The report also recommended that local conservation advisory councils, municipalities, gardeners, landscapers, surveyors, and construction crews all be educated about the vine. Finally, it encouraged volunteer weed-pulling groups as well as the consideration of such alternative remedies as herbicides, incineration, spraying, and grazing. As a result, twenty thousand dollars was put into a contingency fund to monitor the spread of the vine and to provide advice to property owners and municipalities. As McCabe pointed out, "We got support from both parties. Education and knowledge are the best way to deal with the problem."

What was clear three years later, though, was the difficulty of implementing such a broad strategy. The efforts of community activists, conservation advisory council members, and environmental educators to organize neighborhood weed-pulling events flagged when confronted by convictions about the rights of individual property owners. "When one half of the population lets you on their property, and the other half doesn't, it doesn't do any good," Oes-trike says. The county executive's refusal to apply funds for mitigation further assured the spread of the weed. Skeptical about the weed's containment, Oestrike is also apprehensive about what he views as a growing distrust of the scientific community. Citing creationists who resist the theory of evolution, along with ordinary consumers who dispute the effects of releasing vast amounts of carbon dioxide into the atmosphere, he questions whether there is a long-term systematic campaign to discredit science.

And then there is the very persistence of mile-a-minute itself, now trailing through the fall grasses and wildflowers and lacing itself through the chain-link fence at the edge of the ballpark. Its seeds can survive for at least seven years, and its berries can be spread and scattered with swift efficiency by the flow of the stream and flight of birds; ants have been known to carry the seeds, and when deer eat the vine, viable seeds show up in their scat. The elimination of the mile-a-minute requires a like tenacity, but that's unlikely. Pulling it up manually is labor intensive; herbicides damage the surrounding vegetation, and their use near streams is inadvisable; burning it is impractical. Weevils that have been imported from China to eat the vine have not yet been released in this area of New York; the organized effort necessary to get permits for their release takes time and resources that already-stressed community advocacy groups rarely have. The convergence of factors hindering efforts to eliminate the vine seems to have the resolve of the vine itself.

Those efforts, then, have fallen largely to a handful of individual landowners. Property on the south side of the creek is owned by the Blessed Kateri Tekakwitha Church, named for a seventeenth-century Mohawk-Algonquian woman whose conversion to Christianity led to exile from her community. Built in 2007, the gargantuan stone edifice is the largest Catholic Church in the county and capable of seating about a thousand parishioners. Kateri Tekakwitha, whose name translates to "she who puts things in order," was venerated as a saint in 2012, and efforts for her canonization had profiled her as a possible patron saint for ecologists and environmentalists. It makes sense, then, that the church's monsignor, Desmond O'Connor, has tried himself to eliminate the vine from the trees and the fields around the church often used by kids. O'Connor grew up with his grandparents in the Catskills, and his abiding empathy for the region's landscape and waterways is rooted in those childhood experiences. The day I visit him in his parish office, he takes delight in recalling his grandparents' boardinghouse and dairy farm. "I was in 4H," he says, "and we had goats, pigs, sheep, chickens. There were cherry, pear, and apple trees. Everything we ate was right there." Because mowing the vine prevents it from growing tall enough to produce the berries that contain the seeds, the monsignor brush hogs the fields around the church himself several times a season; the drenching rains of spring 2011 prevented him from taking his tractor into the meadow. "It was so wet this year," he says with genuine regret. "I'm behind."

Judith Willsey, the church's neighbor to the north, is as diligent. A middle-aged woman now, she lives in the 1750s house in which she was raised, and her sense of ownership runs deep. For years, she operated a Christmas tree farm, and she became aware of the weed when mowing the fields around the trees in 2007. She acknowledges her role as a temporary custodian of her land. "I'm an Earth

Day person," she tells me one morning when we are sitting on her patio. "Stream walks are a good idea. But how do you motivate and enlist people?" Willsey cites inconsistencies in how ecology is practiced between private landowners and municipalities; the community's collective eagerness to reconfigure the floodplain for baseball fields was in stark contrast to its reluctance to confront the rampant vine and distressed creek. It's a "numbers game," she says, and one that puts a burden on individual property owners. "I pull a lot of it up by hand," she adds, "but there's too much to get a real handle on." She is dismissive of the efforts of a local representative from the Department of Environmental Conservation who advocated the use a small, torchlike device to burn the weed patch by patch. "I'd still be out there doing it now," she sighs.

Her frustration notwithstanding, in the summer of 2011, five years after the vine's discovery at the creek, the presence of mile-a-minute seemed diminished. Willsey continued to mow at the edge of her property, but she saw less of it. And the monsignor told me, "I get out on the tractor, but I have to say I am amazed at how much less there is now. Last year we cut everything right up to the creek, and there were a few spots, but it was nothing like it was the year before." If the vine's proliferation has momentarily subsided, it is difficult to know exactly why. It may be some chemistry of soil and climate or some other characteristic of habitat that remains unknown. Like other invasive weeds, the vine doesn't always advance at the predicted speed. It can occupy prime real estate, then creep into more marginal conditions and dryer soils where it moves less quickly. Oestrike speculated that Japanese beetles, "who like to eat the stuff," may possibly have made a dent in it. Or it may be the prolonged dry spells that recent summers have brought, he said; the vine likes wet conditions and grows most after periods of heavy rainfall.

By the middle of September, though, in the wake of a week of drenching rains from what was left of Hurricane Irene and tropical storm Lee, the vine has reasserted itself on both the north and south sides of the creek, tangled with the nettles and milkweed, goldenrod and jewelweed. Tiny, immature green berries that are more typical of July than September attest to new growth. Dense mats of the vine have covered the shrubs that line the creek, crawl along several small willow saplings, and carpet the trunk of a larger sycamore. While the plants may die off in another few weeks after the first frost, the seeds will last. The berries have natural buoyancy, and the stream, ordinarily shallow and braided, has been reconfigured by the rains into a single, deep, fast flow that is likely to carry them to new areas downriver.

The artist William Christenberry has documented the extreme outcome of a similar vine in a series of photographs that reads as some kind of futuristic allegory of the way nature is capable of devouring what we put in its path. Taken over a period of years, the pictures record the manner in which kudzu vine left unchecked enveloped a small house in Tuscaloosa County, Alabama, the geometry of the small wood cabin giving way gradually to the freak organic topiary of the vine. As it is slowly smothered by leaves, by time, by inevitability, the house morphs into a hybrid structure— part manmade, part natural; partly growing, partly decaying; partly about architecture, partly about landscape, until in the end, it just seems to be an image of some primal struggle between the conflicting impulses of human and plant life.

Eliminating the mile-a-minute vine here altogether would have to have been a larger collective effort among community activists, volunteers, landowners, and legislators, to name only some, and indeed, such a consensus seemed out of reach. And it occurs to me that this may be a time to reexamine what we mean by "collective." Online communities have redefined our capabilities as stewards with their ability to chart the migration of butterflies, follow the flight of raptors, map the moon. Yet this may be one of those occasions when stewardship efforts depend on following more traditional routes of door-to-door, face-to-face, hands-on collaborations; and when we need to consider community not as some digital collective but as a congregation that is defined by geographic proximity and shared civic concerns.

Mile-a-minute vine reminds us what we know in our heart of hearts: despite our instant messages, five-second boot-up times, accelerated lives, when events happen too quickly, it is no good. There is no certainty about how or when the first berries and seeds arrived. The vine did not cause this place to be ruined, nor is the vine a direct result of the impaired stream. Rather, these factors all seem to aid and abet one another in that improvisational and simultaneous manner with which ruin so often arrives. A construction truck arrives with seeds jammed into its tire treads, a bird shits, road salt drains into the streambed, and six new houses are built, necessitating that twelve acres of land be cleared of trees and their roots that absorb water. And then the creek bed becomes degraded, its impairment following a path not unlike that of the vine, insistent, mindless, accidental. It is undone the way so many other things in life are undone, not through choice or design or intent but through negligence and a deficit of care.

The human capacity to tolerate disruption seems nearly limitless. It is like the telephone ringing or the dog barking or the thunder clapping when we are asleep, and then, instead of being awakened by the intrusion, we find a way to fold it into the narrative of the dream; although insulted by some unexpected and jarring interruption, the dream finds some way to go on, accepting and integrating the new assault.

"Science is not perfect," Oestrike had said that day at the creek. "It never claimed to be. But some people need certainty. They need facts. And the facts are always changing. Science is just the best available knowledge we have at one time." That last sentence comes strangely close to the way I have always defined poetry, which is that it is the most we know about the world at any given moment. At times such as this, it is easy to believe that while scientists and the rest of us work, think, imagine, and conjecture in entirely different spheres of experience, it is possible for these to converge.

Jackson Creek and the woods on either side of it seem to mark a boundary between two iconic centers for American community. With its scoreboard, luminescent Coke machine, and festive red, white, and blue bunting hung from its chain-link fences, the ball fields are emblematic of small-town American social life. And certainly the church on the other side of the creek speaks to a different, but no less ingrained, center for our collective well-being. On the day I visit, a group of boys at the church are involved in a project for the scouts. Moving rocks and turning dirt, they are constructing an installation of the stations of the cross, the Christian passage of suffering, contemplation, and deliverance. And I think of the very different progression that has taken place only a few hundred feet away from aspen to birch to willow, where deliverance has yet to take place.

Though with different language and inflection, both the ball fields and the church speak to those ways in which collective enterprise is part of our national profile. We are brought together by many things, but shared faith and playing ball on a Saturday morning are both rituals integral to the weave of our social fabric. It seems hard to understand, then, why some similar common endeavor cannot be applied to the tangle of weed and berries choking the woods that lie between these two foundations of American life. Or why we cannot devise some like social initiative to restore the degraded creek, removing the stranglehold of the vine that runs through it.

11

Insects in the Ash Trees

The natural world and the measures we take to defend it tend to be informed by practical knowledge, by research, classifications, identity, and known facts. The study of ecology is the study of the world that is, not the world that can be imagined. So what to make of it when the world outside your door suddenly takes on the sheen of the fantastic?

Such was the question in June 2011 when I first noticed the purple boxes hanging from the limbs of trees every few miles along the edge of the road running down the valley. Secured with a length of bright yellow twine, they were wedge-shaped and seemed to float in the breeze, exiles, perhaps, from some mass fleet of children's kites or some freak cousin to the Mylar balloons caught in branches and phone lines everywhere, a signal that a rogue block party was about to take place. Like other objects one might hang in a tree—Christmas ornaments, paper lanterns, or a piñata stuffed with sweets—they couldn't help but convey a celebratory feeling on a summer afternoon.

But it was not a party, of course, and bright yellow tags attached to the twine explained that the traps had been hung by the New York State Department of Environmental Conservation as part of its program to monitor the presence of the emerald ash borer, a slender, metallic green beetle currently eating its way through American ash trees, thus far destroying more than sixty million of them. Sixteen species make up the eight billion ash trees in North America, and none are resistant to the ash borers, which feed on the leaves of the tree, devouring the edges in small, irregular patches. The females settle in and bore through the crevices of the bark, laying their eggs in the vascular tissue beneath the bark in summer. Once hatched, the larvae, white, inch-long worms, chew and tunnel their way through the tissue, consuming it and creating circuitous galleries of infection that prevent the host tree from getting nutrients and water. The mortality rate of the trees occupied by the emerald ash borer is 100 percent.

The insect itself doesn't travel vast distances, but like so many nonnative species, it's an intrepid hitchhiker. A native of China, it was first detected in Michigan in 2002, likely having made its way there on shipping pallets. Although loggers now tend to be scrupulous about the wood they transport, the firewood used by campers is often not kiln-dried and can carry the insects any distance a camper might choose to go. By 2009, the emerald ash borer had arrived in New York State, and by the summer of 2011, despite quarantines on firewood, dense colonies had been discovered in counties just west of the Hudson River. Diseased trees begin to thin at their crowns, their branches dying first. Wilting leaves, fresh new shoots growing erratically from the main trunk of the tree, and, sometimes, traces of woodpeckers that happily feed on the larvae can all signal the emerald ash borer. When the adult ash borers leave the tree, they create a small, D-shaped hole in the trunk, but these tiny pictographs of departure can be easy to miss. Often, the infection sets in with few visible symptoms, and it is possible for the beetles to thrive in ash trees for up to two years without detection. It is enough time for the trees to be destroyed.

About 7 percent of the trees in New York are ash, so their disappearance would change irrevocably the texture and character of the forests. But even beyond forest health and diversity, the economic impact of such a loss is substantial. The timber of the ash is used for tools, sports equipment, furniture finishes, and flooring. Ashes line the avenues of urban America, not simply casting their gracious shade across city streets but also absorbing carbon dioxide and providing a carbon sink. They contribute to the comfort of city streets in less quantifiable ways as well, cooling the area around them, reducing wind, and absorbing air pollution. To lose the ash trees is to lose a cherished American resource.

Although the emerald ash borers had been documented at the United States Military Academy at West Point, on the west bank of the Hudson River, they had not yet been found east of the river, the gateway to all of New England in the summer of 2011. The traps, then, were used as a detection device most of all. Teams from the Department of Environmental Conservation put up the 300 or 350 traps in Dutchess County in late spring and early summer. A month or so later, they checked the traps for the presence of the borers, put out additional lures, and made sure the traps were intact, neither dislodged by weather nor removed by residents.

Jerry Carlson, a research scientist with the Department of Environmental Conservation's forest health program in Albany, was overseeing the detection program, and he had put me in contact with the team downstate checking the traps. My hosts on this July afternoon were interns from the State University of New York's College of Environmental Science and Forestry in Syracuse: Blaine Ellis, a graduate student in forestry resources management, and Amy Chianucci, an undergraduate whose major is conservation biology. They suggested I meet them at one of the first sites they'd be checking, the fabled Vanderbilt estate in Hyde Park overlooking the Hudson River. As celebrated as the Gilded Age palace designed by McKim, Mead and White in 1896 are the expansive gardens and grounds that surround it; the collection of specimen trees includes ginkgoes, hemlocks, blue spruces, towering beeches, Japanese and sugar maples, flowering dogwoods, and ash trees as well.

The trap had been hung, though, on a smaller ash in a more

secluded grove near the river. Smaller, but not lesser. Though perhaps not a formal landmark in a renowned American garden, this one was gracious in its own right, with a generous crown and shining oval leaflets that caught a thousand little spoonfuls of light from the river and splashed them back out into the afternoon. The branches of the ash are opposite, growing directly across from each other, and it was easy to see this tree on the riverbank as a kind of living, breathing model of equanimity. Blaine unknotted the yellow twine wound around the trunk that secures the trap and gently lowered it, and I saw what a simple construction it was: a piece of corrugated purple plastic folded into three parts and held together with plastic twist ties and a metal spreader at the top. About a foot wide and two feet long, its vibrant hue had been specified on the speculation that light waves at that far end of the spectrum are more visible to insects.

Hanging from the spreader inside the trap were two small bags of lures containing the pheromones hexanol and manuka oil, both of which mimic ash leaf volatile compounds and attract the beetles. The scent of the lures begins with the faint aroma of olive oil (ash trees are in the olive family), but it's stronger, mustier, and more pungent, and its intensity in the little pouches had a faintly chemical overlay. Once the scent has enticed the insects to the trap, they're there to stay. A film of nontoxic glue has been applied to the textured exterior surface of the prism, so that when insects alight on the box, they are affixed to it; even the name of the glue seems to be consistent with the vivid terminology that suggests both the insect and the means to combat it have been dreamed up by J. K Rowling: the purple prism trap for the emerald ash borer has been coated with Tree Tanglefoot Pest Barrier. I think back to the simple foam noodles, silver tape, and buckets of the weed pull in the river, amazed anew at how the tools we use to address calamities of nature so often recall the objects of play.

Now Blaine and Amy examined the archive of insects that had collected on the prism's surface. Although the emerald ash borer did not happen to be among them, Amy, poised at some intersection of expectation and apprehension, only sighed and added, "It's just a matter of time." The larger specimen trees up by the mansion could be treated individually for infestation, but there are no known treatments that can be safely applied to the vast forests of more ordinary ash trees. For now, Amy and Blaine just added fresh lure bags and hoisted the trap back to its upper limb. He secured the yellow twine with a slipknot around the tree where it would remain until September, when the trap would be checked a second time and removed.

What is happening in the forests of the northeastern United States may not be so much about what is arriving but how quickly it is arriving. A common response to nonnative species is simply that they will find accommodation, that nature will take its course, and that the resolute newcomer will initiate a rebalanced ecosystem. But it's the rate of acceleration that is the problem, suggests Gary Lovett, an ecologist at the Cary Institute who studies forest ecosystems. Lovett is studying the hemlock woolly adelgid, an insect now decimating the hemlocks in some southeastern and mid-Atlantic states, and slowly moving northward; it is now in parts of sixteen states from Maine to Georgia, with impact most severe in Virginia, New Jersey, Pennsylvania, and Connecticut. The insect cannot withstand very cold temperatures for prolonged periods, so it has not yet found the entire Northeast an accommodating habitat. But as the temperature rises, that is likely to change.

There has not been time to draw conclusions about the long-term cycles of nonnative species, Lovett says. "What's different is how fast it is happening. I don't like it. I don't want to lose all these species." Finding accommodation and balance with a blight that arrives in an instant might take a millennium. And hemlock forests could recover, but it might take a thousand years. Although studying the composition of forests is his life's work, he adds, "I can't say what our forests will look like a century from now. Too many things are changing too quickly to take the view that we should just allow all these invasive species to come in, and let nature deal with them. We have no idea what new species will be coming in tomorrow. We are in uncharted territory." Asked about how language around nonnative plants can echo inflammatory rhetoric about immigrant populations, he says, "The metaphor with immigrants is only a metaphor. It's a different thing. We know that some invasive species can do immense damage in a short period of time. If we can identify them and keep them out, we are better off."[1]

I spent the afternoon following Blaine and Amy as they visited nine more sites. Earlier in the season, other members of the Department of Environmental Conservation team had hung traps following a grid system of roughly one-and-a-half by one-and-a-half square miles. Because the county was now a high-risk zone, several traps were hung in each square of the grid. The tree specified for a trap almost always grew within the ten-foot right-of-way from the center of the road, the distance within which the researchers can install traps without securing permissions from private landowners; with resources already stretched, the administrative costs of contacting so many private property owners might easily derail the entire effort.

Whenever possible, the purple prisms had been hung in trees on open ground and in the sun. In some cases, though, that hadn't been possible, and locating the ash trees, particularly those growing in dense thickets along the side of the road, could be a trick in itself. In this part of the county, the back roads between the various sites are often little more than antique cow paths that thread through old farms, past remnants of vintage stone barns and historic cemeteries. At one site, when the GPS unit in the silver Jeep Cherokee Blaine and Amy were driving suddenly issued an unexpected spasm of instructions, the entire enterprise seemed like something between a scavenger hunt and some real-world board game. It was an impression only reinforced by the fact that the traps are often called "Barney Traps," a reference to the TV dinosaur whose hue they share.

The suggestion of play was reiterated at the conclusion of each site visit when Blaine rehung the traps after he had checked them. Attaching an arborist's small orange sandbag to the end of the yellow twine, he tossed it overhead, looping the twine twice around the overhead branch the trap was to hang from, then drawing the twine down again, looping it and knotting it around the trunk of the tree. Although the traps may have appeared to be floating from the limb overhead, they were, in fact, securely attached; were they to be dislodged by wind or weather, they might scrape along the trunk and branches, allowing the glue to be rubbed off. If locating the sites, tossing the orange sandbag, and looping the yellow twine all parodied the simple moves of some outdoor game, it was perversely apt to the afflicted ash tree: its wood is light, strong, elastic, and ideal for the manufacture of baseball bats. Before the advent of metals and composites, it was the favored material for hockey sticks, tennis rackets, swings, and playground equipment. Louisville Slugger harvests northern white ash trees for its major league baseball bats near the New York–Pennsylvania border, and the manufacturer's website informs readers it is working closely now with the USDA and state department of agriculture in Pennsylvania to monitor the insect's presence; that said, the company is already testing other species of trees.

Still a forester in training, Blaine worked with a natural courtesy toward the woods; an intuitive custodian, he avoided stepping on a rambling old stone wall "because it's just too pretty," and if he had had to hack at the bitterweed or poison ivy that had tangled itself around the twine, he did so without inflicting damage on the tree. He and Amy have developed that brand of camaraderie, some combination of humor, simpatico, patience, and resignation, that often springs up between people who are trying to take care of something or someone in the face of overwhelming odds.

As we drove up to one residential site, Blaine suddenly caught his breath. "Oh my God, is that a chestnut?" Amy stopped just short of rolling her eyes and murmured, "It's his favorite tree. Every time we see an American chestnut, we have to pull over." "That's because they're all dying," Blaine responded, sprinting from the car to the lawn where, just for one brief moment, he manages to put aside his reservations about trespassing on private property without permission. It turned out that the tree was not the landmark of American forestry, after all, but a chestnut oak. Still, in an afternoon spent looking for something we hope not to find, it somehow seemed consistent to have the illusory moment of imagining you *have* found something you cherish. Blaine may be a man after the facts, but his scientific detachment dissolved during a brief contemplation of the American chestnut. "I even tried growing one," he said, "but it didn't work out." "That must have been hard for you," said Amy, teasing and sympathetic, before both turned their attention to the trap on the adjacent ash tree.

Up close, the graphic language rendered on each panel of the prism resembled a gorgeous abstract painting. For all its precise geometry of form, the compositions found on its exterior were more improvisational. The kingdom of insects had been flattened out on a dazzling gestural canvas, a kinetic calligraphy of tiny flies and moths, mosquitoes, wasps, a daddy longlegs, tiny beetles, and here and there a lightning bug, thrown in with bits of leaf and branch, pollen and seed, and the samara, the little wing-shaped seeds of the ash tree, all as though the artist Cy Twombly had set up his studio here in the woods of Dutchess County to work up these precise yet inscrutable compositions.

The three-sided purple prism trap had all the clarity its name suggests. Its accidental composition, both random and utterly informative, was all about what comes next. Given the data it contained, its surface resembled a pictograph representing the past as well as the future. And while it happened to be swinging from the bough of an ash tree on this July afternoon, it could just as easily have been hanging on walls of the conceptual art museum Dia: Beacon, only a few miles to the west. People talk about the convergence of the arts and sciences, but I promise you, *promise you,* that if you have a purple prism trap hanging anywhere nearby, take a good look, because that conjunction is happening right there below the branch of an ash tree.

Yet if I was inclined to see the surface of the trap as some ravishing expressionist canvas, Blaine and Amy studied it with a different absorption, scrutinizing its surface, gently picking at it with their dissecting tweezers, decoding and deciphering all its raw data with the concentration of Talmudic scholars. I saw now that checking the traps was a short sojourn to a separate universe, one that engaged the senses as well as the intellect. In the sultry heat of the July afternoon, the oily aroma of the lures was oppressive, and the sticky surface of the prism along with its lurid color all managed to create a deranged little ecosphere of their own.

Blaine lifted a tiny, desiccated click beetle off the surface with his tweezers; its slender shape, size, and dark, gleaming shell all resembled those of the emerald ash borer, but when he compared it with the borer he carried in a small bottle for just this purpose, it was clear that the latter had a more pronounced head and larger, bulbous eyes. Should there be any further question, emerald ash borers also have purple wings concealed just beneath their metallic green shells.

Blaine had already turned his attention back to the surface. "This has all kinds of little things on it, a tiny little beelike thing, and eggs," and sure enough, along with mayflies, leafhoppers, feathery threads of milkweed, a fragment of cicada shell, and a tiny green grasshopper, an infinitesimal row of white dots neatly lined up as testimony to some small insect's instinct for continuance. But again, the emerald ash borer remained absent. "I don't expect to find it," said Amy, "but I don't not expect not to find it either," using that obscure language of hesitation, uncertainty, and double negatives that human beings so often turn to when they know it's only a matter of time before undesirable information turns up.

Or perhaps it was just her growing fluency with ambiguity. If I had come to all of these outings and excursions with the premise that scientists work strictly with the facts of the observed world, I was only partially correct. Over the recent months and years, I had often wondered what, if any, was the essential difference in the way the scientific community and the rest of us view life here on earth. It occurred to me now that in their persistent search for dependable data, most scientists live with doubt far more comfortably than the rest of us; they are not just conversant with the unknown but friendly, gracious, even intimate.

From time to time throughout the afternoon, local residents driving by slowed down to ask what Blaine and Amy were doing. It was an interest in the purple prisms that reflected their success in a secondary endeavor: although the size, shape, aroma, and color of the traps have been determined by their value in attracting a particular insect, the same characteristics also elicit human curiosity. Such attention doesn't come often. In matters of stewardship, insects tend to occupy the lowest areas in the hierarchy of human interest. People respond first to the plight of animals, sentient creatures with eyes that see and hearts that beat, creatures capable of evoking human empathy. Next come plants, visible to us, and treasured, whether it is for the food they provide, the beauty they exhibit, the shade of trees, or the aroma of flowers. But the infinitesimal beings of the insect world, barely visible, tend to become abstract, and our response to them is usually indifference. "We would like to use citizen scientists so people could learn about the emerald ash borer, but we don't have the funding to set up the program," Jerry Carlson had said to me on the phone a week earlier. "We could use people to collect insects, monitor and check traps, make sure they're not vandalized, cut down, shot down, or dislodged and damaged by weather." Potentially, the Q&A by the side of the road can be a precedent to the knowledge and attentiveness unlikely to come to this insect population by other, more common routes.

The improvisational roadside briefing may also be one more instance of that emerging affiliation between scientists and stakeholders that William Schlesinger has called "translational ecology." Certainly the curiosity of local residents, sometimes infused with skepticism and suspicion, and Blaine's clear answers, seems like a small step toward building such an alliance.

The last tree of the day was a scrawny little ash we drove by once before we had to circle back to find it. Whoever had hung the trap earlier in the season had overtied it, wrapping the twine around the trunk several times, then knotting it needlessly and elaborately. Blaine groaned as he undid the trap and finally lowered it, but there was something in the excessive efforts of his predecessor that seemed familiar; from time to time we probably all operate on that flawed equation which allows us to believe that our most profuse efforts, even when utterly pointless, will somehow help to ensure the desired outcome.

"I just love ash trees," Amy said, more to herself than to either of us. "Now when I drive around, and I see an ash tree, I notice them, I point them out. They are just so beautiful." Is it easier to love something you know or something you don't know? You could probably make a good argument for both perspectives. It is easy to love what you don't know, when the object of affection remains remote, when it resonates with mystery and wonder, shaped and colored and altogether formed by your own desires, expectations, illusions. But at those times when we allow our esteem to be extended to those things we have made an effort to know well, a different kind of imagination comes into play, and I would suspect it is a manner of ardor that is, in the end, easier to sustain.

But it was the end of the afternoon by then, and a red-tailed hawk sliced a low arc across the sky. Blaine and Amy had checked thirteen traps without finding evidence of the emerald ash borers. "Just because you don't see them doesn't mean they're not there," Amy said with a sigh. She knew they were not far off, just as she knew there was a world of difference between what is unseen and what is absent. "I don't want to find it. But I would still like to have it in my collection," she said. Not yet twenty-one, she had already made full acquaintance with that awful anomaly of searching for something you hope you will never find.

When I had first seen the purple prisms hanging from the trees in early summer, it had been hard not to see something festive, celebratory in them; later, those notions were dispelled. Now, at the end of summer, as they were removed from the ash trees, what came to mind most was the archetypal wish tree that exists in so many cultures, such as those gnarled pines in Japanese temple courtyards where pilgrims knot their small notes; or an oak tree in Scotland I have heard of where votive coins are hammered into the trunk; or wedding trees in Holland on whose branches guests attach their manifesto of hopes for the marriage couple.

The conversations we have with the natural world are likely to be as varied and wide-ranging as those we have with one another. Yet I suspect there is something in the way we speak to trees that often has the tone of supplication. It seems to come naturally to us to see the tree as that form in nature that can reflect and express our deepest wishes, as a symbol of continuance and endurance. Perhaps it is some primal instinct to imagine that a tree in full leaf can contain and carry those things we desire. Yet what was hanging in the trees in Dutchess County that summer signaled a hope for the *tree,* the symbol itself under duress, and I wonder if that shift is some sign of the dire place where we have found ourselves. While the message of the purple prism traps is surely more complex and layered than that of a small coin or handwritten wedding wish, it, too, seems to carry on this tradition, some iteration of our pagan longing to be returned to a more ordered universe, some far-fetched petition for an accord with the world around us.

Postscript: In March 2012, evidence of the emerald ash borer was found in three ash trees in northern Dutchess County, eliminating any lingering hope that the Hudson River had been a geographic barrier to the insect. Early detection may help to prevent its rapid spread, and individual afflicted trees can be treated. Elsewhere, landowners can plan for new species to replace the loss of ash trees. Effective ways with which to arrest the beetle's advance or to treat large areas of distressed trees are not yet known.

12
Eagles on the Shore

Scorekeeping, statistics, and noting how things add up or don't come naturally to most human beings. I suspect most of us are equipped with some instinctive method of weights and measures. It's what we use to track our losses and gains, and probably also what registers the events of the natural world when the math is off, whether it is a profusion of weeds, a diminishing number of bats, or even just a torrential rain. Perhaps, too, these exercises in tallying eels or counting herring or measuring beds of wild celery are all ways to compile numbers, to shore up this interior numbering system that gives us an essential sense of earthly order.

Still, much of such accounting is done loosely; events and observations are jotted down in memory in a kind of intuitive record keeping that we do only because it seems natural to do so. In *The Universal History of Numbers,* George Ifrah notes that "logic was not the guiding light of the history of number-systems. They were invented and developed in response to the concerns of accountants, but also of priests, astronomers, and astrologers, and only in the last instance in response to the needs of mathematicians."[1] There are times, too, when the concerns of this constituency may inform the more disciplined accounting of science or when one sphere of information intersects with the other. When aquatic geologist and limnologist John J. Magnuson was researching freeze and breakup dates of ice in the northern hemisphere, he looked to records kept by Shinto priests in a shrine alongside Japan's Lake Suwa. For over five centuries, the holy men had documented those dates on which the lake iced over and when cracks appeared in the ice, the latter an indication that God had set foot there.

There is something reassuring in knowing that our efforts to bring structure to the more transient events of the natural world have their own tradition and that numbers can provide a framework for our conjectures, a system to which we can fasten nothing more than our hopes, guesses, and speculations. Certainly that is the way I have been keeping count of eagles as they've made their return to the Hudson Valley.

By 1970, they'd become nearly extinct. The fish and waterfowl that were their prey species had absorbed high levels of DDT from pesticides washing into the river. When ingested by eagles, the chemicals thinned the raptors' eggshells, making them incapable of surviving incubation. The raptors' diminishing numbers were further assured by habitat loss brought about by a growing human population and the penchant to create appealing river vistas by cutting down the tall sycamores and cottonwoods favored by eagles for nesting. A restoration program was initiated in the 1970s when the endangered species unit of the Department of Environmental Conservation trapped Alaskan eaglets for release in the Hudson Valley. The ban on DDT, habitat protection, and education all helped them to flourish, and by 1989, ten breeding pairs of eagles were nesting in New York. Today, the Hudson Valley is one of the largest wintering areas for the birds in the lower forty-eight states. More than 170 pairs nest in New York State, with some twenty-five or thirty of those nests in the Hudson River watershed alone.

I've used my own improvisational eagle count to track that recovery for myself. The hunting club that owns much of the land in our valley raises pheasants and ducks, and its ponds on either side of the road have made it a popular habitat for both the year-round eagle population and the winter migrants. For years, then, I have made it a habit to scan the winter sky as I drive down the valley, searching for the eagles among the chaotic cross-hatching of locust branches or in the upper limbs of old oaks. Still, I am always surprised when I actually see one. It is an assault on expectation, a confirmation that something wild can still happen, as on the November day when an adult bald eagle sailed down in front of my car, then lifted slowly toward the marsh just to the north, a sudden flash of the seven-foot wingspan, the radiant arc of its white tail, the snowy dome, all gone as quickly as they came. When eagles soar, their wings spread flat in a horizontal plane, unlike the "V" formation made by some other raptors, and there is something in that straight line that always resets the day's axis, bisecting the day into before and after.

Or the more subdued appearance on a drizzly January day when the melting snow, the brown grass beneath it matted with rotting leaves, and wet bark all conspired to keep the day in sepia. As though attired in camouflage for just this stippled day, a juvenile bald rested on the upper limbs of an oak at the edge of the pond, the mottled hues of its feathers in chromatic accord with the shades of the morning.

Or the frigid February morning a couple of winters ago, ten or fifteen degrees below zero at 7:00, the coldest it had been in a couple of years. A car had struck a deer in front of our house the previous night, a Saturday, and the road crew wouldn't come until Monday to pick it up. All day Sunday, an immature bald eagle pecked at its entrails. We live near a crossroads, so cars generally slow down when they approach, but if one went by fast, the juvenile would sail up to the limb of a locust tree in the marsh, watch, wait, and return to its bloody banquet in the road only after the car had passed. I know a cardinal law of eagle watching is to stay away, keep a distance, don't intrude, don't frighten the bird to move and needlessly burn calories. Eagles have an "alert distance" of about eight hundred feet, although their eyesight is so keen that they probably see you far sooner. More important, they have a "flight distance" of about four hundred feet. These are averages, and some birds will show more tolerance than others; but breaching the distance and startling the bird into flight causes it to expend energy, which, in winter, is both wasteful to the eagle and dangerous.[2] But the carcass and bird were practically in my front yard, and too many rules had already been broken—the speed of the cars, the deer in the road, the eagle's proximity to the house—so it was hard to feel guilty standing on my own front porch watching the bird just thirty feet away.

Or the first really warm spring day in mid-March when my friend Jane and I took a walk down to her pond. For all the signs of growth—the first early sprouts, the buds on the trees, the slight shift in ground color from taupe to celery—there were signs of disruption and upheaval as well—a clamoring of crows, a small burst of turkey down on the ground, the odd tuft of pelt, tiny explosions of fur and feather, pheasant and bluebird both, and random bits of mayhem everywhere. And then we saw them, in that way that the majestic sometimes comes upon you accidentally, two, four, then all five juvenile bald eagles, a pageant in the sky over the pond. The trees in the marsh and woodland were still bare, their muted branches and trunks laying out a certain structure, but the sky scavengers with their giant wingspans seemed to carve out some parallel network in the sky, as though giving the air itself a dimension I hadn't known about. "I can just feel all the green underneath about to come up," Jane said, but what's happening in the sky was out of scale with even that expectation.

All of these sightings make for a kind of vague and inexact tallying, but it has nothing to do with the deliberate count that is made when accuracy is essential, a matter of discipline and necessity. Real mathematicians even draw a distinction between accuracy and precision: the first has to do with the closeness of what is being measured to the given standard, while precision has more to do with how measurements, when repeated, show the same results. A measurement system can be precise but not accurate, or accurate but not precise. I know that my own interior score sheet has a long way to go before it will ever recognize such differences, but still, it is my hope this winter to witness a more solid and reliable accounting system. And certainly the Department of Environmental Conservation's Annual Mid-Winter Bald Eagle Census is one that can be put to practical use. Made in the lower forty-eight states along 740 established survey routes, with data taken on foot or by car, boat, helicopter, and plane, the census is an effort to create an index of wintering eagle populations; to identify, manage, and protect their habitats; and to learn about their migratory pathways and movement patterns.

Yet when I left the house at 7:30 on an early January morning, a full moon slipping between low clouds in the western sky, it was already clear that its name was a misnomer. It was twenty degrees, and by the time I got to the Hudson a half hour later, the mercury had risen to the low thirties, and it was another unseasonably mild morning. Midwinter seemed a good half season or so away, and the river's wide sheen of gray was running clear without any trace of ice. White's Marina in New Hamburg looked like some improvisational sculpture park, its dozens of fishing boats, pleasure cruisers, and yachts all tightly wrapped with blue tarps, a boatyard that might have been reconfigured by Christo. Or perhaps, disguised and distorted, the warped watercraft were just monuments to the warp in season.

I was here to meet Tom Lake, a naturalist for the New York State Department of Environmental Conservation's Hudson River Estuary Program, an anthropology teacher at Dutchess Community

College in Poughkeepsie, and the editor of the *Hudson River Alma-nac*. He had agreed to let me tag along today as he surveys twenty-nine sites along both sides of a twenty-five-mile section of Hudson River shoreline for the census count. Along with date, time, and site, the survey form asks for start point and endpoint, the particular body of water, survey method, whether the site was a roost, and whether the eagles are mature or immature. It asks for temperature, weather conditions, wind, cloud cover, and ice. It is detailed, specific, and comprehensive, and it has no use for guesswork.

But all we could see from the New Hamburg marina were a scattering of gulls and a single common merganser flying low upstream. Lake speculated that most of the eagles were still up on Lake Champlain or the Saint Lawrence River. "There is no urgency for them to be here," he said. "Eagles have been coming south for fifteen thousand years. To the estuary for twelve thousand years. Ducks and fish are their primary food source." Two different populations of eagles use the Hudson River as habitat during the winter. Along with the year-round resident population is the vast migrant population of wintering eagles that fly south from northern New England, Ontario, and the Canadian Maritimes when waterways to the north have frozen over, sealing off their favored food sources. Though it is hard to be specific about their numbers, some 250 eagles come from the north to areas along the Hudson and Delaware Rivers. If the Delaware and Catskill reservoirs are locked up with ice, there may be as many as 200 migrant eagles along this area of the Hudson. This year, though, warm temperatures have left open water in the north, allowing the eagles to remain in their breeding habitats.

Scientists and naturalists often tend toward a reserved manner, but Lake doesn't hesitate to express his excitement, and his speech patterns reflect this as much as what he actually says. When he is conveying information of which he has a great deal, he tends to speak rapidly, but when he is telling you about something that is strange or unknown or mysterious, his voice slows down. He gestures not just with his hands but with his whole arms. And as with many people who are serious and knowledgeable about their field of study, he is completely attuned to its comedic aspects, such as the time an eagle flying overhead dropped a small sturgeon on his neighbor's head. These things happen. And he doesn't hesitate to reference human behavior. When discussing the poor manners, distractedness, and self-absorption of immature male eagles, his speculations go directly to teenage boys. And when the subject is the human fascination with eagles, he says with resignation, "We love them. They don't love us. It's unrequited love." Then he sighs. "Too much love. It just isn't going to work out." Tom Lake's observations are so astute, his sense of history is so broad, his eye so discerning, that such conjectures can only be a valid addendum to his knowledge base.

We headed a few miles south, but nothing much was going on there either. Across the river in Newburgh are the Roseton and Danskammer power generating stations. When the river is frozen, the plants' warm water discharge makes for a good area of open water for the eagles to find the fish they are seeking. Add to that a few cottonwoods standing between the two plants that offer good nesting sites and convenient feeding perches, and there are some frigid winters when the ice floes on this area of the river become small continents of eagles. But this was not one of them. The raptors from the north hadn't yet found any good reason to migrate so far south this winter. "They're finding their food upriver," Tom said.

"There is no ice on the river up to the foothills of the Adirondacks."
A couple of years earlier, the helicopter count along the Hudson
River and southeast New York had documented a record count of
277 birds, 142 adults and 135 immatures. But as Lake has noted
in the *Almanac,* "Counting eagles can be imprecise, and totals must
be compiled carefully. Along the same one-to-two mile reach of the
river, exactly one hour apart, the number of birds changed from four
to ten back to three." If counting eagles is part of some effort to
quantify what is wild in our lives—and surely it is—perhaps such
elusiveness is inevitable.

Measuring ice on the river is as ephemeral an exercise. Com-
plete ice records need to factor in the duration of ice, its thickness,
and the percentage of its cover across the body of water—all of which,
of course, are ever changing. On the Hudson River, such information is even more fugitive due to the number of its tributaries and
the salt content of the lower river. Still, ice measurements can provide a reliable record of climate change, as was revealed in 2000 by
Magnuson's research. Researching 150 years of ice cover on lakes
and rivers in the northern hemisphere, he found that freeze now
occurs 5.8 days per 100 years later, while ice breakup occurs 6.5
days per 100 years earlier, averages that indicate warming temperatures.[3] Although similar long-term records of ice on the Hudson
are not available, US Coast Guard cutters typically begin breaking
ice on the river in mid-December, finishing the season at the end
of March. In the winter of 2010–2011, for example, the ice first
appeared on the river on December 16, and ice remained on northern areas of the river until March 10. This year, though, no ice will
be recorded on the river until January 16. By the eighteenth, the
river will be open water again. A day later, half an inch of ice will

appear again in some areas, and by January 24, one to two inches of ice will form. But by January 29, the river will be open water again and stay as such through the rest of the winter.[4]

Our next stop was across the river in Balmville, just above the city of Newburgh. Lake laid out a chart of the river on the hood of his car to give me a primer on eagle habitats: north to northwesterly winds prevail on the river, so eagles prefer to put in at south-facing peninsulas that protect them. They are looking for ways to minimize energy loss; the less wind and hard weather they're exposed to, the less time they'll have to spend hunting. He focused his binoculars on the far shore at Brockway, the site of an old brickyard just north of Beacon. It was hard to tell from this distance, and it could have been a black walnut or locust tree, but a single bald eagle was perched on one of its upper branches. I squinted, readjusted my binoculars, scanned the riverbank. It always amazes me how far the other side of the river is. After a minute or so, I made out the tiny white blur. Here we are, I thought, looking at America's totemic raptor on the banks of one of its grandest rivers, but what I could see was just a bit of pale smudge on the far edge of gray water. If this isn't the syntax of the momentous, I don't know what is. Almost always, it seems to back into our lives quietly.

Eagles are also known as a sentinel species, which are defined as those fish, birds, and animals at the top of the food chain whose health inevitably reflects the health of the environment around them; if a sentinel species is stressed by toxins in its diet, for example, it can serve as a model for how these stresses can influence human well-being. Eagles, ospreys, great white sharks, tigers—such sentinel species are large, few in number, prone to extinction, rare, and, often, creatures of myth that appeal to the imagination. Such nomenclature seems especially appropriate for such a raptor. With eyes that can see both ahead and to the sides, it can spot a fish in the water from hundreds of feet. And while a solitary eagle scanning the river from the upper limb of a tulip poplar may well be an indicator of ecosystem health, there is something in its watchful stance and bearing that allows us to imagine it an agent for a different kind of well-being. It probably comes naturally to most human beings to engage in some kind of occult association with the animal world. Myths and fables of every continent and culture are given to tracking such imagined alliances. Whether it is in strength, endurance, knowledge, beauty, or some other quality we find insufficient in ourselves, humans have always managed to find some psychic empowerment in the animal kingdom. And if we have distanced ourselves from those inclinations today, eagles are the exception. We continue to find in them some spirit of ascendance. We recognize them for being a sentinel species. All the more reason to track their numbers.

Our next stop was the Newburgh Boat Launch, a couple of miles downriver. It was 35 degrees by now, and the sky was clearing. An enormous common merganser drifted placidly among the gulls. There was a black scoter, a sea duck, probably in migration. And across the river, about a mile wide here, two adult bald eagles had settled in a sycamore tree. Again, Lake had picked them off instantly, but by the time I have adjusted my binoculars, one is gone. The other remained, nothing but a white spot on the far shore. I remembered how the bats in the locust tree were no more than a faint humming sound and the herring at Hunter's Brook nothing but quick movement, a shift of light in the water. How easy it is to miss what we are after.

We drive a few miles south to Plum Point but are greeted only by a few gulls. "In real winter weather, you'd see more immatures," Lake said. "The ones we are seeing today are probably only local birds." Juvenile bald eagles tend to congregate, and if the migrant birds were this far south, this is where you might be likely to find them. We look up across the river to Denning's Point, a thin peninsula that stretches south into the river from Beacon. The river is shallow there and, at low tide, offers an extravagant buffet to the eagles.

A bit farther south in Cornwall, a flock of Canada geese fly low over the water. Gulls resting on the rotting debris and logs left in this shallow area of the river by Hurricane Irene flap and call. A few buffleheads drift along the river, some mallards, too. Eagles mate for life, and this is the time of year they are refurbishing their nests, Lake told me. The two we saw up at the Newburgh launch may well be a pair that has been roosting here for four or five years. We looked out across the water. Eagles will cause pandemonium when they land on the same log in river with gulls, he said, but sometimes the gulls know they're not hungry and will agreeably share the same spot. "How do they know the eagles aren't hungry?" he asked rhetorically. "I don't know. No one knows that." But it was clear that such confoundment pleases him. For everything he knows about the eagles on the Hudson River, he seemed happy that the raptors and gulls have kept *some* information to themselves.

The woods across the river were dense with black locust, white oak, sycamore, and cottonwood, that is to say, good eagle trees, which Lake defines as being on or near the river, often in sheltered locations, out of the prevailing wind, with a sunny exposure and a good water view. They are likely to have a large, open canopy along with large, horizontal limbs that make good feeding perches. Many of these trees have been cut down in recent decades to open up broad river vistas. ("Viewsheds," Lake snorts. "Frederic Church knew about viewsheds, too. He just didn't call them that.") "The formula for an eagle tree is easy in, easy out," Lake says, but today the eagle trees remain vacant. It's ten o'clock now. The clouds have thinned, revealing a pale blue sky. It's forty degrees and feels like an early April morning. He's headed to a meeting in Westchester, but he'll visit more sites on his way back up. By the end of the day, though he will have traveled fifty miles of shoreline, his count will stay at those we've seen this morning.

I know what an eagle tree looks like. The previous April, my friend Polly had taken me to nearby Bowdoin Park on the eastern bank of the Hudson River, where a pair of eagles was nesting in the crotch of a towering tulip poplar. The Department of Environmental Conservation had been monitoring the nest, confirming that two eggs had hatched on or around March 31; a couple of weeks later, two nestlings were verified after the adults were seen bringing food to the nest. The afternoon we're there, it's the first hot day of the spring, well into the mid-eighties. The female eagle was in her nest, an expansive ramshackle bowl of branches and twigs, an odd pine bough or two woven into the constriction at haphazard angles. The eaglets remained out of sight, though Polly tells me she had glimpsed bits of down the day before. At rest, the female is vigilant, her snowy head in clear profile as she shifts her stare from the river just to the west down to the park below where kids are playing and a few hikers are walking along the trails. Eagles are disturbed by jet skis, motorbikes, outboard motors, loud vehicles, and loud people, but what's happening in the park on a spring afternoon, the laughter of children and footfall of hikers, is a more quiet manner of play and seems to intrude little on the plane of activity in the branches overhead.

Moments later, the branches sway, the air everywhere seems to move, and the male comes in for a landing on an adjacent tulip tree. For all his weight and size, he lands lightly on the upper branches and settles in, pecking at himself, ruffling his feathers, his gaze ranging to the nest, to the river, and out across to the banks on the opposite side. The delicate tracery of the winter branches, barely even in bud, seems a fragile edifice to support his full weight, but out here on a spring afternoon, a different system of balances seems to be in play.

Not long after, I learn of a parallel view to the eagles. Midway across the country, on April 2, 3, and 6, three eaglets were hatched at a fish hatchery on the banks of the Trout Run in Decorah, Iowa. The nest, some six feet wide and four feet deep, is eighty feet high in a cottonwood tree, and two cameras owned and operated by the Raptor Resource Project have been installed several feet above to provide continuous real-time footage of what is happening in the nest. When I log on in mid-April, it is snowing in Decorah, and the female eagle is in the foreground, working to reshape the nest, the wind howling, the slate gray river streaming behind her. She reconfigures the nest, picks up bits of twig and branch, repairing it, reshaping it. Both her position on the screen and her efforts to keep the eaglets warm and protected make it difficult to see them. In the lower section of the frame are the pop-up ads for yogurt, cereal, time-shares, reverse mortgages, gold coins, thermal imaging cameras, reduced airline tickets. Forty-three million viewers have logged on so far.

The following day is warmer, the snow has melted, the trees are blowing less, and the mother is moving around, less preoccupied with keeping the nestlings covered and warm. She sits back, allows them to be exposed, rearranges a few twigs, and the three eaglets, little tufts of gray down are moving about, nipping at one another. By midafternoon, the sun seems brighter still, and two eaglets are clumsily tangled up in each other, pecking at bits of straw in the nest. A gust of wind blows, and one burrows beneath the mother. The other just seems to collapse in on itself. She appears to be keeping the third, smallest eaglet beneath her to keep it warm. A mourning dove lets out its soft cry over the river. Such is the minutiae of life in the cottonwood tree. For the minute, detailed view, the electronic surrogate is irresistible.

Over the next two months, I visit these two nests intermittently, the tulip tree at the park and the cottonwood I can reach more easily by logging on at my computer. The chicks were all hatched within days of one another, and it is impossible not to consider their parallel lives. Storms in Iowa sometimes cause the lens to ice over so all that is visible is a sheet of glistening silver. Most of the time, though, the view offers a mesmerizing proximity. The male is in the foreground, his tail feathers consuming the screen, their scalloped patterning precisely defined, their iridescence ranging from a brown sheen and dark gray to russet to a deep purple, nuances of color shifting imperceptibly, the tip of each feather catching the light to become a pale tawny shade. The feathers lie flat; a moment later, they ruffle. And then a lift in the air, the shudder of leaves, and he's gone. Or the female sorting things out in the nest. The wind has reconfigured her living quarters, and a slender branch in full leaf has twined itself across and through the nest. She pecks at it repeatedly, trying to put it in place as any fastidious homemaker might. That's not the only thing in her way; one of the eaglets is spreading its small wings in the nest, then gets up on its spindly feet and totters around for a minute or two before settling back in with its siblings. I try to articulate why it is so hard to stop watching this and can come up with nothing better than the meditative character of these small lives at the top of a tall tree at a time when life elsewhere moves too quickly. I wonder if any of the eighty-five thousand other viewers who are logged on at the same time could say anything more definitive.

Times passes in a different way at the park. One drizzly April day, the nest appears empty, but within minutes the female sails in from the river. She stands at the nest, dropping bits of fish she has brought with her. From time to time it is possible to see a small, dark head pop up, gawk at the wide world below, then retreat. The male is nowhere in sight. A few afternoons later, the nestlings are visible to the bare eye, reaching and stretching for bits of food. An elderly couple is watching them with binoculars, and the woman tells me how she watched as the male and female added to the nest. She lives minutes away and comes often to check on them. They're big, they're just getting so big, so quickly, she says. "We went away for three weeks on vacation, and the first thing I wanted to do when we got home was come over here." The female lifts suddenly and positions herself on an upper limb of the tulip tree. The eaglets are awake now and reach their necks up to her, but she is scanning the fields below and the river.

By mid-May, the buds in the tulip tree are opening, and pink flowers frame the vast, patchwork nest. The profiles in the nest seem unreasonably large. Behind the lace of early spring leaves is the flutter of dark wings, the eaglets lurching around, flapping their wings, strange little dinosaur birds with their ungainly limbs and awkward jerks. A park ranger in a green John Deere golf cart rolls up, and I ask him what he has had to do to safeguard the nest. "Nothing," he tells. "Really, nothing. The Department of Environmental Conservation comes to check on them, and from time to time, if people are standing right beneath the nest, we ask them to move, but that's about it." Mike lives nearby, and he tells me the nest is four years old, though nothing had ever hatched there before. "But I knew things were different this year," he tells me. "I watched him bring in bits of moss to make it soft inside. Now I watch to see what he brings in." He takes out his cell phone to show me a picture of the male with a striped bass hanging from its talons. "Another time, I watched him bring in a squirrel," he says, with some combination of awe and amusement that suggests it is completely possible to take custody of things we do not own.

By the end of the month, the tulip tree is in full leaf, its foliage screening the nest. The nestlings have become more venturesome. One is perched close to the edge, seeming to outgrow the nest altogether as it tucks into the adjacent limb to find a new fit with the universe just beyond. A gust of wind rearranges the leaves around it, and suddenly the world is a place of precipices—the nest, the branch, the season, the age of the nestling. Then the female soars in from the river, feeds them what she has brought back, stays for a few minutes, then glides back to the river. The nestling flaps its wings as though testing them, learning perhaps how feathers can catch the air, then hops around in the nest in its gangly fashion, chirping its awful chortle, then settles in again, a portrait of impatience, hunger, boredom, restlessness, that combination of mundane incentives that impel so many living things toward grand gestures and decisions.

Life in Decorah is no less perilous. The camera pans out over the limb of the cottonwood where two fledglings perch. One opens its wings wide, gives a little jump, and for a moment hovers. The third fledgling remains in the nest, scratching itself. A moment or so later, all three scuttle and lurch along the branch. There is a great flapping of wings, but they go nowhere and, instead, remain lined up on their limb in a constellation of anticipation. The airfare ads bring a certain hilarity to the proceedings.

Visible only intermittently, what happens in the nest in the park is largely a matter of conjecture. The spectacle of an adult soaring in from the river with a herring in its beak unfolds suddenly and unpredictably against the wide sky and river. But the view of the Decorah eagles is constant, framed, contained, every detail captured; it can be prolonged and meditative, simultaneously intimate and remote in that way that the camera allows. Maybe this is just another iteration of something I already know, the difference between the screen truth and the air truth, but both ways of watching take place in real time, and if one is more real than the other, I couldn't say which. It is a convergence of images that has already been cut and pasted into my memory in some composite form that may speak to how we experience nature in the twenty-first century.

A month after the census, I saw Lake again at a midwinter eagle watch he offers upriver. It was early February, but again, unseasonably mild in the low fifties. During the ice-over of the intervening weeks since the census count, migrating eagles from the north had come south. Although the ice had vanished by now, the visitors seem to have stayed. Upriver and across, a pair of adult courting bald eagles was perched on the upper limb of a hardwood, staring across the water in a remote portrait of repose. Twenty minutes, twenty-five, thirty, the two remained on their upper limb, motionless but for the occasional flutter of feather or turn of head. A mammoth barge, the *Clipper,* with red and blue stripes and yellow stacks, drifted down river, but they remained unfazed by the industrial pageant below. How long will they stay there, I wondered aloud. Lake looked at me with incredulity. "They're in *love,*" he said with exasperation. "You can't quantify love. This is *bliss.* You don't *know* how long. They could be there for hours."

Just downstream, another pair of bald eagles was soaring and wheeling in aerials above the gray stone towers of an old seminary, the stasis of the edifice underscoring their lightness of being. They arced out of view. And then, midriver, a single eagle alighted on a conifer tree on a small island. It paused, then a moment later glided along the surface of the water, flapped its great wings, touched down on the water. Its talons flashed and gripped something along the way before it settled on an outcropping of rocks, hacking at whatever piece of fish it had picked up.

One river. Forty minutes. Five eagles. One day's numbers for restoration. I think of all the different ways we have of tallying those things we care about and am astonished anew by our proficiency in calculating the evanescence of what surrounds us. That we have found ways to count birds and measure ice makes me think it could be possible to decipher the increasingly uncertain world that we are bound to face. But my level of confidence is not 95 percent.

Maybe we bring a particular ingenuity to the process when these things could be lost so easily. How do you measure what you love? Maybe it is just one of those equations that involve both the necessary and the impossible. I would like to think that when the accounting system addresses recovery rather than loss, our mathematic imagination comes alive. Philosophers differ on whether we have any innate capacity to understand zero and infinity, but my own suspicion is that it is the kind of knowledge that comes and goes. There are days when the words "never," "always," "forever" have an absolute clarity to them, and other days those same words are completely beyond our reach and comprehension. Perhaps, then, all of our assorted accounting systems simply reflect a curiosity about that distance between nothing and everything, between never and forever; perhaps they are just the result of some inevitable conjecture we all have about our own position in the continuum. "What is man in nature?" Pascal asked. "Nothing in relation to the infinite, everything in relation to nothing, a mean between nothing and everything." I'm not sure whether his equation is accurate or precise, but this afternoon, at least, it sounds true.[5]

Things move on without us, and I wonder if there is a way to count ahead in time. The eagle on the rock has finished its meal and scans the river south, east. Then it lifts, sails north just above the gray sequined surface of the water, and vanishes up the river's long hallway.

Epilogue

Xavier de Maistre was an eighteenth-century aristocrat and writer most known for his travelogue of the voyage around his bedroom. Wearing his pajamas, Maistre did his best to examine his sofa, windowsill, and bedpost as though a tourist, bringing a traveler's fresh perspective and insights to the view of his rugs, armchair, and prints. While his work parodied the great travel narratives of his contemporaries who were prone to allegorical rhapsody, it also speaks to the possibilities of bringing imagination to our consideration of those rooms and landscapes so familiar to us.

And it occurs to me that all these naturalists are engaged in some equivalent effort. Perhaps in some similar way, they have come to think of their own towns and communities as their rooms. Dan Smiley spent his entire life recording everything he encountered on his sky island, a single quartz ridge in Ulster County, New York. Thoreau found a sanctuary and a universe at Walden Pond, noting in his essay "Walking" that "there is in fact a sort of harmony discoverable between the capabilities of the landscape within a circle of ten miles' radius, or the limits of an afternoon walk, and the threescore years and ten of human life. It will never become quite familiar to you." Thoreau's "grandest epiphanies always came locally," says Simon Schama, and cites a journal entry by Thoreau written at his pond in twilight: "The moon traveling over the ribbed bottom—and feeling that nothing but the wildest imagination could conceive of the manner of life we are living. Nature is a wizard. The Concord nights are stranger than the Arabian nights."[1]

Staying close to home and looking closer to the ground, into water, or at animal life, it is possible to experience a similar curiosity, a similar appetite for information, a similar astonishment at the unexpected answers. It is late December 2011, and not unlike Maistre examining his sofa and paintings, I scan the water's surface on the pond down the road to see if there is ice cover yet. When we moved here all those years ago, it was a marsh, but the steady efforts of the beavers have long since made it a pond. IceWatch USA is a data collection program that invites volunteers to observe and record the formation of ice on their local waterways, along with air temperature, snow depth, rainfall, and presence of wildlife. Seasonal changes in freezing and thawing affect what else happens in the ecosystem, and the database that gathers this information from different locations will be used to help determine continuing effects of climate change.

Volunteers are asked to select one small part of a body of water, part of a river or an inlet in a lake, and to record ice-on and ice-off days. Early morning December 3 was when I first saw a faint scrim of ice on the water. Night-time temperatures were in the low teens, and it was only twenty degrees by nine that morning. By midday, the ice was gone. It reappeared on the morning of the fourth, but warmer temperatures later in the month prevented the water from freezing. Now, at month's end, the pond has yet to ice over. If there is any quantifiable outcome of my time spent looking for the bats, the egg masses in the vernal pools, or the eels in the stream, it likely has to do with accuracy in searching out the minutiae; for attending to the particulars; and for suspecting any quick conclusions I might be tempted to draw. More rigor for fact, more room for doubt. Knowledge and skepticism converge and inform each other; the strength of their alliance is not something I had understood before.

What I do not doubt is that looking across the water for ice can constitute travel. Like any other tourist stunned into silence by the panorama from the summit or blinking in disbelief at the strange artifacts in the souk, I find it easy to be astonished by what I am seeing.

The value of such local efforts on conservation is beginning to find consensus. This was one of the conclusions at a conference about citizen science held in New York City in spring 2011. Cosponsored by the American Museum of Natural History's Center for Biodiversity and Conservation, the Audubon Society, and the Cornell Lab of Ornithology, the gathering of sixty practitioners, including scientists, teachers, government workers, and representatives from assorted nonprofit agencies, addressed how scientific research and conservation efforts might better collaborate. Among the points made in the workshop summary was that "through the sharing of stories and experiences, we heard how effective collaborations tie conservation efforts to larger issues such as water and environmental policy, land use, sociology, and socio-economics. Glimpses into existing projects illustrated the potential for public participation in scientific research to be mutually beneficial for scientists and the public, enabling scientists to gather large quantities of data that would otherwise be cost-prohibitive, and participants to connect and engage with the management of resources that directly impact their lives." The summary went on to note the common inclination of people from all walks of life to notice seasonal phenomena without even being aware of it and that such observations were an opportunity "to move the public from being observers to being actors."[2]

The tone of reserve in this final note is somewhat puzzling, because it is, *of course,* increasingly difficult for such seasonal phenomena to escape the notice of the general public. While rising levels of carbon dioxide—invisible, colorless, odorless—may escape our notice entirely, extreme flooding, heat waves, wildfires, tornado outbreaks, massive droughts, and the effects these have on seasonal phenomena generally *do* capture our attention. In years past, the three or four weather catastrophes that occurred each year in this country cost about a billion dollars each. In 2011, such disasters have cost more than fifty billion dollars. It is inevitable that we will make observations of events and phenomena that will surprise, disturb, and otherwise discomfort us. Going from being an observer to becoming an actor is likely to be sudden, spontaneous, and involuntary, as on the morning a friend discovered an infestation of armyworms nibbling on the chard, leek, kale, arugula, lettuce, and pretty much everything else in her greenhouse. More common to southern states, the worms may have been driven north by the winds from multiple summer storms. Their proliferation in the Northeast could be a product of dry weather followed by heavy rain. But other than the infestation of the caterpillars, none of this was certain.

When does a fact acquire meaning?

"We are changing the large-scale properties of the atmosphere—we know that beyond a shadow of a doubt," Benjamin Santer, a climate scientist at the Lawrence Livermore National Laboratory told the *New York Times* in December 2011. "You can't engage in this planetary experiment—warming the surface, warming the atmosphere, moistening the atmosphere—and have no impact on the frequency and duration of extreme events."[3] A majority of climate scientists have now reached a consensus that the increasing number of extreme weather events is the by-product of a warming planet, and that rising temperatures accelerate the global water cycle, resulting in more storms. Yet the particulars of how things happen, when, and why remain less clear. All the more urgent, then, is the call for more research and a broader engagement of the scientific and nonscientific community alike in addressing such questions as how to reduce energy consumption, limit carbon emissions, manage natural resources, and develop renewable energy resources.

The air temperature at sunup this morning was thirty-eight degrees. The wild grasses at the banks are flattened and crumpled from pounding rains the night before, and the haphazard geometry of dried cattails and phragmites seem to offer some equation for the rough edges of the season. It is winter, but barely. Beneath the dark surface of the water lies the vibrant green tracery made by a bed of watercress. It is almost New Year's, and ice-on remains a matter of days away. Such is the manner of one thing becoming another. But watching for water to turn to ice and ice to water is probably the least of it; the transformations to come are likely to do so in a more extreme, unexpected, and unknown manner. Erik Kiviat's words come back to me: "There is no end of the story in ecology. There are only probable outcomes."

If it was Thoreau's conviction some century and a half ago that it may take the wildest imagination to grasp the manner of life we are living, some further reach of human ingenuity might be needed to comprehend the speed and character with which the natural world is changing today. How to grasp fully the effects of a warming global temperature may require some originality of thought and action that is not yet clear to us.

Routine and revelation have always been steadfast allies, and it is my suspicion that such an imagination arrives most frequently through measured work, and that local epiphanies are a common product of rote discipline. At a time when apocalyptic rhetoric so often colors the conversation around changing global ecologies, it may be useful to consider what are sometimes called repetitive motion tasks. My guess is that whatever degree of wildness happens to be required, it is likely to come through the conjugations of ordinary action, through that impeccable eye and certain hand that come from the repetition of familiar gestures—whether it is weeding the river of water chestnuts every July, counting the eels coming into the tributary every Monday in the spring of your sophomore year in high school, or taking the temperature at the same time and same place every morning of your adult life.

APPENDIX

Because it is natural for people to forge connections with the places in which they live, most efforts toward stewardship are likely to begin locally, whether it is in documenting birds, assessing vernal pools, or monitoring air quality. Such grassroots endeavors often begin by word of mouth, kids and parents meeting at schools, or neighbors encountering each other at the local recycling center. Other efforts may begin on social networking sites. But in almost all cases, digital networking can expedite collecting, organizing, and sharing data. Mobile phone applications facilitate the identification and monitoring of everything from birds and leaves to night sky constellations; as the *Wall Street Journal* has noted, "Smartphones are the butterfly nets of the 21st century" (Jennifer Valentino-DeVries, "App Watch: Mapping Nature on Your Smartphone," *Digits* [blog], February 28, 2011). The social dynamic these sites and apps promote nurtures a collective effort, a sense of collaboration and community that is particular to these times. I have made an effort here to catalog a variety of citizen science programs—and their assorted websites and apps—that look to the work of volunteers whose individual observations can be submitted to an organizing agency to become part of a larger data set.

Some of these are long-standing programs such as the Christmas Bird Count. Most are more recent. Cycles of the natural world can take an instant or a generation, and the timetables of stewardship are no less varied— a single day, week, season, or year. Some of these are barely participatory, requiring nothing more than spare computer processing power on participants' personal computers; crowdsourced science can be long on the crowd and short on the science. But others ask for volunteers to analyze digital data. And others still ask participants for "boots-on" observations—that is, first-person, on-site documentation of what they see, hear, and witness themselves. These are strategies for attending to things that may be as close and familiar as what is stored in your own refrigerator or the nine-spotted ladybug nibbling on a zinnia outside your window. Or as distant and remote as a crater of the moon. Many of these programs address traditional information gathering, whether it is about weather patterns, wildlife population, or discoveries in outer space. Others work to track and document nonnative species and their effect on local environments. And in recent years, other protocols have also been established, out of necessity, to help participants assist in monitoring distressed environments, whether it is the contamination of ground, air, or water, the pollution of noise or light.

I have not included regional programs, focusing instead on those with a national or international reach, but many of these do include local affiliates. Protocols may vary from region to region, state to state. Bucket Brigade, for example, is a national organization that enables citizen activists to monitor air quality, but its local entities address the specifics of particular landscapes, land uses, industries, and toxins. Likewise, the national eagle census count takes place on different days in different states.

Nor have I included time-specific programs necessitated by such unanticipated disasters as the 2010 oil spill in the Gulf of Mexico or the earthquake and tsunami that disabled the Fukushima Daiichi nuclear power plant. After the 2010 Deepwater Horizon catastrophe, Gulf Oil Spill Tracker allowed local residents to monitor and share information about the leaked oil, whether on water, on land, on wildlife, or in the air; and in 2011, RDTN.org was a site that raised funds and purchased and delivered Geiger counters to residents of Fukushima, allowing them to measure the levels of radiation for themselves. In both cases, data collection by local residents was instrumental in assessing damage. In the face of government inaction and corporate denial, it is not an exaggeration to suggest that such citizen activism has become integral to how a community or an entire region may recover from catastrophic assault. But because their time frame and duration are, by definition, specific and limited, I have chosen not to include these initiatives here.

Some of these programs are geared toward students, some toward adults. Many depend on the efforts of both. It did not seem essential to distinguish between the two here. This selection is broad but by no means comprehensive. Still, my hope is that it will provide a place to begin to find the ground truth, the sky truth, the air truth, the water truth.

Adventurers and Scientists for Conservation: This nonprofit puts together partnerships between the scientific and research community and adventure athletes—mountain climbers, hikers, bikers, skiers, snowshoers, rafters, kayakers, and gliders—whose far-flung areas of activity can offer the opportunity to collect data otherwise hard to come by on species ranging from ice worms and zooplankton to grizzly bears, wolverines, and whales. www.adventureand science.org

Belly Button BioDiversity Project: Closer to home, this project proposes that the navel could be one of the last biological frontiers and invites participants to identify and learn about the bacteria that this well-protected refuge of the human body hosts. www .wildlifeofyourbody.org

BioBlitz: A rigorous biological survey carried out by scientists and volunteers to identify all species in a particular area within a limited period of time for the purpose of raising awareness of biodiversity. www.nationalgeographic.com/explorers/projects/ bioblitz/

Bucket Brigade: In an effort to reduce pollution, improve safety, and enforce environmental legislation, this simple air-monitoring system enclosed in a canister enables residents in industrial areas or near refineries, chemical plants, or other sites that may compromise air quality to collect data about what they are breathing. www .bucketbrigade.net

Celebrate Urban Birds: An accessible bird observation kit enables urban birders of all ages to compile and submit data to scientists at the Cornell Laboratory of Ornithology. www.birds.cornell.edu/ urbanbirds

Christmas Bird Count: Inaugurated in 1900 by the Audubon Society, America's longest running wildlife census looks to the efforts of tens of thousands of volunteers who document regional bird populations every December. www.Birds.audubon.org/ christmas-bird-count

Citizen Weather Observer Program: This public-private partnership collects data gathered by more than eight thousand participants; makes it available to weather services and homeland security; and provides feedback to contributors to ensure quality of information. www.wxqa.com

CoCoRAHS: Weather observers of all ages contribute to Community Collaborative Rain, Hail and Snow Network to compile a comprehensive national record of daily precipitation levels. www .cocorahs.org

Digital Fishers: Volunteers use their own computers to analyze fifteen-second deep-sea videos recorded by underwater cameras from the Neptune Canada observatory to help identify ocean biodiversity and species behavior. www.digitalfishers.net

EarthWatch: Established in 1971, this international nonprofit organization invites volunteers to work as research assistants with scientists in projects that range from monitoring climate change in the Arctic to studying leatherback sea turtles in Costa Rica to tracking moths in southern England. www.earthwatch.org

eBird: Established in 2002 by the Cornell Lab of Ornithology and the National Audubon Society, the online checklist program allows professional and amateur birders to report, access, and share information about bird species. www.ebird.org

Einstein@Home: Initiated in 2005 as part of World Year of Physics, this program allows volunteers to download software to process data to identify and analyze unknown continual gravitational-wave sources using data collected by LIGO (Laser Interferometer Gravitational-Wave Observatory) and then sends the data back to a central server. www.physicscentral.com/experiment/einsteinathome/

ExCiteS: Extreme Citizen Science is an interdisciplinary research group that includes anthropologists, computer scientists, and engineers and differentiates itself from other endeavors of data monitoring, collection, and sharing by providing a system enabling diverse communities, both rural and urban, to establish their own community-specific programs with tools that can be easily used. www.ucl.ac.uk/excites

Firefly Watch: As numbers of fireflies decline, the program allows volunteers to gather information about geographic distribution and activity of local fireflies and submit their information to become part of a larger database. www.mos.org/fireflywatch

Folding at Home Project: This distributed computing project uses participants' personal computers to simulate protein folding in research for such diseases as Alzheimer's, Parkinson's, autism, and cancer. www.folding.stanford.edu/

FrogWatch USA: Volunteers identify, count, and track breeding and habitats of frog and toad species to learn about wetlands and to

help scientists researching amphibian management. www.aza.org/
frogwatch/

Galaxy Zoo: Volunteers view and help to classify hundreds of
thousands of galaxy images using data collected and archived by
NASA's Hubble Space Telescope. www.zooniverse.org

Global Amphibian BioBlitz: As some one-third of the world's
amphibians face extinction due to such factors as climate change
and habitat loss, this offers amateurs a platform to observe, photograph, and record local amphibian populations. www.iNaturalist
.org

Global Coral Reef Monitoring Network: Established for the
observation, management, and conservation of the world's coral
reefs, this also has a reef check program that trains divers on how
to observe and record trends in the health of coral reefs. www
.gcrmn.org

Global Reptile BioBlitz: Amateur observations of reptiles assist
scientists in mapping and conserving habitats of more than nine
thousand recognized reptile species. www.iNaturalist.org

The Gravestone Project: Volunteers check marble gravestones at
graveyards worldwide to measure the weathering rate of the marble,
an indicator of changes in rainfall acidity. www.goearthtrek.com/
gravestone

Great Sunflower Project: Volunteers in urban, suburban, and rural
areas who plant sunflowers can then watch and count the bees these
attract to assist scientists in understanding declining populations of
native bees and honeybees and the effect on pollination of garden
plants, crops, and wild plants. www.greatsunflower.org

IceWatch USA: Volunteers choose a location—river, lake, bay—and
record and report observations on ice formation and ice melt to

assist scientists studying changes in weather patterns and their impact on wildlife. www.natureabounds.org

Jelly Watch: Volunteers contribute to a long-term data set by reporting information about jellyfish, squid, red tides, and other unusual marine organisms, conditions and occurrences to help biologists better understand oceans. www.Jellywatch.org

Journey North: To help learn how monarch butterflies respond to changing climates and seasons, the program enables volunteers to track them each fall and spring as they migrate between North America and Mexico. www.learner.org

Lost Ladybug Project: Noting time, date, weather, and habitat, volunteers find, collect, and photograph ladybugs to assist entomologists in understanding changes in species habitat and distribution. www.lostladybug.org

Mercury Poisoning Project: The use of elemental mercury in some Latino and Caribbean homes in a ritualistic attempt to repel evil spirits can have long-term adverse health effects on residents. While no coordinated studies have been done to date on the outcome of such practices, this project is an effort to begin gathering such information. www.mercurypoisoningproject.org

Midwinter Bald Eagle Survey: This program invites volunteers in the lower forty-eight states to work with federal and state agencies, along with scientists, to observe, count, and record bald eagles during the first two weeks of January. To document species recovery, 740 established survey routes are followed in cars, planes, boats, and helicopters. Dates vary by state. http://corpslakes.usace.army.mil/employees/bird/midwinter.cfm

The Milky Way Project: To help map star formation, participants identify and chart bubbles, knots, and star clusters using data from the Galactic Legacy Infrared Mid-Plane Survey Extraordinaire (GLIMPSE) and the Multiband Imaging Photometer for Spitzer Galactic Plane Survey (MIPSGAL). www.milkywayproject.org

The Monarch Program: This volunteer tagging program advances knowledge about monarch migration populations and patterns and how they may be affected by climate trends and human encroachment. www.monarchprogram.org

Moon Zoo: Volunteers review images made by NASA's Lunar Reconnaissance Orbiter to count and document the craters of the moon, along with any other distinguishing physical characteristics of the moon's surface. www.Moonzoo.org

Nature Mapping: A partnership of communities, businesses, schools, and other agencies to collect data about wildlife, water, plants, habitats, and climate for purposes of education and inventory of biodiversity. www.naturemappingfoundation.org

NestWatch: Volunteers in this monitoring project from the Cornell Lab of Ornithology observe and record data, including location, habitat, species, number of eggs, number of young, and number of fledglings. Information is submitted to an online database and compiled with other observations in an effort to better understand the behavior of North American bird species and to help further their reproductive success. www.nestwatch.org

NetQuakes: This monitoring program allows participants to install digital seismographs in private homes, businesses, schools, and public buildings to record and transmit data online to the USGS after an earthquake. http://earthquake.usgs.gov/monitoring/netquakes/

NoiseTube: This mobile phone application enables participants to use their phones as environmental sensors by measuring noise levels in urban areas, then sharing this information to compile a noise pollution map. www.NoiseTube.net

North American Breeding Bird Survey: Conducted across randomly determined roadside routes in the continental United States and southern Canada, this monitoring program tracks more than four hundred species of birds, recording their population and distribution. www.pwrc.usgs.gov/bbs

Old Weather: Distinguished by its view to historical data, this program asks volunteers to recover observations made by Royal Navy ships in the early twentieth century, information that is then used to help scientists build new models for weather and climate. www.oldweather.org

Project Noah: This mobile phone application—the acronym stands for "networked organisms and habitats"—enables users to explore, document, and share wildlife data. www.projectnoah .org

Project FeederWatch: A project of the Cornell Lab of Ornithology and Bird Studies Canada, the winterlong survey from November until April asks participants to count and report birds seen at feeders, helping scientists to track trends in winter bird populations, as well as their distribution and abundance. www.birds .cornell.edu/pfw/

Project Squirrel: Volunteers in rural, suburban, and urban areas record observations about fox and gray squirrels, which, active year-round, are reliable indicators of local ecologies. www.project squirrel.org

The Public Laboratory for Open Technology and Science: Using do-it-yourself techniques, PLOTS (Public Laboratory for Open Technology and Science) develops and applies open-source tools for environmental investigation, monitoring, and documentation. http://publiclaboratory.org

QuakeCatcher Network: A collaborative initiative to develop the world's largest strong-motion seismic network, the program uses sensors attached to networked laptops and desktops to create a tool for both education and early warning systems. www.qcn@stanford .edu/

SafeCast: This global sensor network allows participants to compile, collect, and share data about local radiation measurements. www .safecast.org

SciSpy: This mobile phone app from the Science Channel enables participants to use their phones as a field kit, allowing them to gather information about plants, insects, birds, and other wildlife and to take photos, upload, find, and share information. www .scispy.com

SETI: Berkeley-based SETI (Search for ExtraTerrestrial Intelligence) uses a network of Internet-connected home computers to search for signs of intelligent life outside Earth. The search system is integrated into a screensaver that is put into use only when the computer is not in use by its owner. www.SETI.org

SkyTruth Alerts: The real-time alert system alerts participants to events such as toxic spills or air pollution damaging land, sea, or air and then allows them to monitor these using an online digital mapping system. Trained volunteers can assist in image processing and analysis and GIS web-based mapping and digital graphics. www.skytruth.org

SOHO Comet Hunting: SOHO (Solar and Heliospheric Observatory) is a joint mission of the European Space Agency and NASA that provides satellite images that can be downloaded and searched by amateur astronomers for evidence of comets, then reported and checked for accuracy. www.cometary.net

Solar Stormwatch: Volunteers use home computing to analyze images tracking explosions on the sun and their subsequent passages across space. www.solarstormwatch.com

Watch the Wild: Volunteers choose a specific location or route, record observations about wildlife, including flowers, trees, plants, insect life, animals, streams, and general weather conditions, and then report these to a central database. www.natureabounds.org

Whale.FM: This joint project of Scientific American and Zooniverse invites volunteers to use spectrogram imagery to help marine researchers catalog the calls of killer and pilot whales that have been collected in a database. www.whale.fm

What's Invasive: The data collection program using smart phone applications allows volunteers to observe, document, and report invasive plants and insects in national parks worldwide. www. whatsinvasive.com

The WildLab: The mobile phone application enables users, largely students between grades five through twelve, to collect GPS-tagged bird sightings that are then sent to the Cornell Lab of Ornithology for research purposes. www.thewildlab.org

Wildlife of Our Homes: This site invites participants to gather samples and share information about the microscopic species—among them bacteria, archaea, protists, pollen, and fungi—with which we share our domestic lives, all the while considering the particulars of their habitats: building materials, carpeting, ventilation, heating, and cooling. www.yourwildlife.org

YardMap Network: This ecological social network from the Cornell Lab of Ornithology allows birders, gardeners, and other online participants to map habitat management and carbon-neutral practices in backyards and parks and share the information on a Google map. www.bird.cornell.edu/citscitoolkit/projects/clo/yardmap

NOTES

Chapter 1. Introduction

Epigraph: Wilson, *Diversity of Life,* 351.

1. Burgess, *Daniel Smiley of Mohonk,* 109.
2. William H. Schlesinger and Jessica Vitale, "Historical Analysis of the Spring Arrival of Migratory Birds to Dutchess County, New York: A 123-Year Record," *Northeastern Naturalist* 18 (2011): 335–346.
3. Abraham J. Miller-Rushing and Richard B. Primack, "Global Warming and Flowering Times in Thoreau's Concord: A Community Perspective," *Ecology* 89 (2008): 332–341.
4. Loren Eiseley, *The Immense Journey* (New York: Vintage Books, 1959), 65.
5. David A. Seekell and Michael L. Pace, "Climate Change Drives Warming in the Hudson River Estuary," *Journal of Environmental Monitoring* 13 (2011): 2321–2327.
6. Schlesinger and Vitale, "Historical Analysis."
7. Julia Frankenstein, "Is GPS All in Our Heads?" *New York Times,* February 5, 2012.
8. Wilson, *Future of Life,* 133.
9. Lake, *Hudson River Almanac,* November 16–18, 2011.
10. E-mail correspondence with Rick Bonney, May 23, 2012.
11. Judith Enck, "Opening Remarks," EPA Citizen Science Workshop, New York, June 19, 2012.

12. Rick Bonney et al., "Public Participation in Scientific Research: Defining the Field and Assessing Its Potential for Informal Science Education," Center for Advancement of Informal Science Education (CAISE), July 2009.

13. Telephone conversation with Rick Bonney, June 20, 2012.

14. Sallie McFague, "New House Rules: Christianity, Economics, and Planetary Living," *Daedalus* 130, no. 4 (2001): 125–140.

15. Peter Preuss, "Empowering Communities—Air Pollution Sensors and Apps," EPA Citizen Science Workshop, New York, June 19, 2012.

16. Terry L. Root et al., "Fingerprints of Global Warming on Wild Animals and Plants," *Nature* 421 (2003): 58.

17. Janis L. Dickinson, Benjamin Zuckerberg, and David N. Bonter, "Citizen Science as an Ecological Research Tool: Challenges and Benefits," *Annual Review of Ecology, Evolution, and Systematics* 41 (2010): 149–172.

18. Wilson, *Diversity of Life,* 351.

19. William H. Schlesinger, "Translational Ecology," *Science* 329 (2010): 609.

20. Gary Lovett, "Climate Impacts on Forests and Ecosystems," Climate Change in the Hudson Valley, conference, Cary Institute, Millbrook, NY, October 22, 2011.

Chapter 2. Bats in the Locust Tree

1. Muir, *Thousand-Mile Walk,* 164.

Chapter 3. Weeds on the River

1. Erik Kiviat, "Under the Spreading Water Chestnut," *News from Hudsonia* 9, no. 1 (1993).

2. David Strayer, interview, "The Roundtable," WAMC, Albany, NY, January 23, 2012.

Chapter 4. Pools in the Spring

1. Damon B. Oscarson and Aram J. K. Calhoun, "Developing Vernal Pool Conservation Plans at the Local Level Using Citizen Scientists," *Wetlands* 27, no. 1 (2007): 80–95.
2. Silvio O. Funtowicz and Jerome Ravetz, "Science for the Post-Normal Age," *Futures,* September 1993, 739–755.
3. Wilson, *Diversity of Life,* 347.

Chapter 5. Ribbons Underwater

1. Chip Brown, "The Future Isn't Futuristic Anymore," *New York Times Magazine,* January 29, 2012, 19.
2. Carol Vogel, "True to His Abstraction," *New York Times,* January 22, 2012.
3. Reinhold Niebuhr, *The Irony of American History* (New York: Scribner Library of Contemporary Classics, 1985), 63.
4. Presentation by Stuart Findlay, Cary Institute of Ecosystem Studies, Millbrook, NY, June 2012.

Chapter 6. Coyotes Across the Clear-Cut

1. Roland Kays, Abigail Curtis, and Jeremy L. Kirchman, "Rapid Adaptive Evolution of Northeastern Coyotes via Hybridization with Wolves," *Biology Letters* 6, no. 1 (2010): 88–93.
2. Conversation with Pam Golben, environmental educator, Hudson Highlands Museum, Cornwall, NY, October 27, 2010.
3. Telephone conversation with Javier Monzon, Department of Ecology and Evolution, Stony Brook, NY, June 8, 2012.
4. Thoreau, *Walden,* 137.
5. Julian Treasure, "The Four Ways Sound Affects Us," Ted Talks, TEDGlobal, July 2009.
6. Heather Wieczorek Hudenko, Daniel J. Decker, and William F.

Simere, "Stakeholder Insights into the Human-Coyote Interface in Westchester County, New York," Human Dimensions Research Unit, Department of Natural Resources, Cornell University, February 2008.

7. Telephone conversation with Monzon, June 8, 2012.

8. Mark E. Weckel et al., "Using Citizen Science to Map Human-Coyote Interaction in Suburban New York, USA," *Journal of Wildlife Management* 74 (2010): 1163–1171.

9. Ibid., 1169.

10. Wilson, *Diversity of Life,* 350–351.

Chapter 7. Herring into the Brook

1. Annie Dillard, "Seeing," in *Pilgrim at Tinker Creek* (New York: HarperPerennial, 1974), 18.

2. Ibid., 20.

3. Jorge Luis Borges, *Dreamtigers,* trans. Mildred Boyer and Harold Morland (Austin: University of Texas Press, 1964), 43.

4. Conversation with Vicky Kelly, Cary Institute of Ecosystem Studies, Millbrook, NY, April 27, 2011.

5. Burroughs, *Locusts and Wild Honey,* 55.

Chapter 8. Loosestrife in the Marsh

1. Daniel Q. Thompson, Ronald L. Stuckey, and Edith B. Thompson, "Spread, Impact, and Control of Purple Loosestrife (*Lythrum salicaria*) in North American Wetlands" (Washington, DC: US Fish and Wildlife Service, 1987).

2. Banu Subramaniam, "The Aliens Have Landed! Reflections on the Rhetoric of Biological Invasions," *Meridians* 2, no. 1 (2001): 34.

3. Larson, *Metaphors for Environmental Sustainability,* 4–6.

4. David Strayer, "Manage Pathways to Block Invasive Species," *Poughkeepsie Journal,* January 31, 2010.

Chapter 9. Eels in the Stream

1. Lori Quillen, "An Interview with Steward Pickett, Urban Ecology Visionary," *Eco-Focus* (Cary Institute of Ecosystem Studies) 6, no. 1 (2012).
2. John Steinbeck, *The Log of the Sea of Cortez* (New York: Penguin Classics, 1995), 92–93, 93–94.
3. Prosek, *Eels,* 9.
4. Ittelson et al., *Introduction to Environmental Psychology,* 51.
5. Ibid., 52.
6. Richard Feynman, "The Value of Science," address to the National Academy of Sciences (Autumn 1955), published in *The Pleasure of Finding Things Out: The Best Short Works of Richard P. Feynman,* ed. Jeffrey Robbins (Cambridge, MA: Perseus, 1999), 146.

Chapter 11. Insects in the Ash Trees

1. Conversation with Gary Lovett, Cary Institute of Ecosystem Studies, Millbrook, NY, April 27, 2011.

Chapter 12. Eagles on the Shore

1. Ifrah, *Universal History of Numbers,* xxi.
2. Lake, *Hudson River Almanac,* January 14, 2011.
3. John J. Magnuson, "Historical Trends in Lake and River Ice Cover in the Northern Hemisphere," *Science* 289 (2000): 1743–1746.
4. Telephone conversation with Kary Moss, chief warrant officer, US Coast Guard, June 28, 2012.
5. Blaise Pascal, *Pensées* [1670], ed. and trans. Roger Ariew (Indianapolis, IN: Hackett, 2004), 59.

Epilogue

1. Henry David Thoreau, "Walking," in *Nature/Walking*, ed. and intro. John Elder (Boston: Beacon Press, 1991), 79–80; Thoreau quoted in Schama, *Landscape and Memory*, 577.

2. Engaging and Learning for Conservation: Public Participation in Scientific Research, April 7–8, 2011, http://www.birds.cornell.edu/citscitoolkit/conference/ppsr2011/.

3. Justin Gillis, "Extreme Year, Few Measures," *New York Times*, December 25, 2011.

SELECTED
BIBLIOGRAPHY

Burgess, Larry E. *Daniel Smiley of Mohonk: A Naturalist's Life*. Fleischmanns, NY: Purple Mountain Press,1996.

Burroughs, John. *Locusts and Wild Honey*. Boston: Houghton, Osgood, 1879.

Del Tredici, Peter. *Wild Urban Plants of the Northeast: A Field Guide*. Ithaca, NY: Comstock, 2010.

Felstiner, John. *Can Poetry Save the Earth? A Field Guide to Nature Poems*. New Haven: Yale University Press, 2009.

Henshaw, Robert E. *Environmental History of the Hudson River*. Albany: State University of New York Press, 2011.

Ifrah, Georges. *The Universal History of Numbers: From Prehistory to the Invention of the Computer*. New York: John Wiley and Sons, 2000.

Ittelson, William H., et al., eds. *An Introduction to Environmental Psychology*. New York: Holt, Rinehart and Winston, 1974.

Lake, Tom, ed. *Hudson River Almanac* [weekly]. New Paltz, NY: Hudson River Estuary Program, New York State Department of Environmental Conservation.

Lannoo, Michael J. *Leopold's Shack and Ricketts Lab: The Emergence of Environmentalism*. Berkeley: University of California Press, 2010.

Larson, Brendon. *Metaphors for Environmental Sustainability: Redefining Our Relationship with Nature*. New Haven: Yale University Press, 2011.

Leopold, Aldo. *A Sand County Almanac and Sketches Here and There.* New York: Oxford University Press, 1949.

Muir, John. *A Thousand-Mile Walk to the Gulf.* New York: Houghton Mifflin, 1916.

Prosek, James. *Eels: An Exploration, from New Zealand to the Sargasso, of the World's Most Mysterious Fish.* New York: HarperCollins, 2010.

Schama, Simon. *Landscape and Memory.* New York: Vintage Books, 1995.

Stanne, Stephen P., Roger G. Panetta, and Brian E. Forist. *The Hudson: An Illustrated Guide to the Living River.* New Brunswick, NJ: Rutgers University Press, 1996.

Strayer, David L. *The Hudson Primer: The Ecology of an Iconic River.* Berkeley: University of California Press, 2012.

Thoreau, Henry D. *Walden: A Fully Annotated Edition.* Edited by Jeffrey S. Cramer. New Haven: Yale University Press, 2004.

Tingay, Ruth E., and Todd E. Katzner, eds. *The Eagle Watchers: Observing and Conserving Raptors Around the World.* Ithaca, NY: Comstock, 2010.

Wilson, Edward O. *The Diversity of Life.* Cambridge, MA: Belknap Press of Harvard University Press, 1992.

———. *The Future of Life.* New York: Alfred K. Knopf, 2002.

INDEX

accuracy, as mathematical term, 187–188
Adirondack State Park, 58
air pollution, 20
alewives, 108, 110, 118
algae blooms, 50–51
"Aliens Have Landed! The" (Subramaniam), 130
American Acclimatization Society, 131–132
American bittern, 124
American chestnut tree, 176–177
American eel, 49, 136–137, 141, 145
American goldfinch, 127
American Museum of Natural History, 205
American woodcock, 11
amphibians, 6, 57, 64, 69, 127
Angevine, Mark, 140–141, 147–148
Annual Mid-Winter Bald Eagle Census, 188, 210
apple trees, 11, 28, 60, 72
Arcade Fire (band), 27
architecture, regional, 13
armyworms, 206–207
ash trees, 67, 169, 171, 180, 182; baseball bats made from wood of, 175–176; potential disappearance of, 170–171; wing-shaped seeds of, 177
Asian longhorned beetle, 131

Asiatic tear-thumb, 157
aspen trees, 167
asters, 67
Atlantic Ocean, 45, 48, 106, 137, 142
attentiveness, 4, 40, 65–66, 120; conditions of visibility and, 112; herring monitoring and, 107; varied circumstances of, 7
Audubon, John James, 22
Audubon Society, 205
autumn, 15, 134

Bachelard, Gaston, 32
bacteria, 58
bald eagle, 8, 81, 185–186, 192
barberry, 159
Bard College, environmental activism at, 59–60
barn swallow, 107
Basho, 16
bass (fish), 8
bats, 8–9, 42, 73, 184; echolocation capacity of, 33; in folklore, 35; fungus disease as threat to, 37–38; longevity of, 34–35; maternal colonies of, 29–30, 38; radio-tagging of, 7, 31; tree roosts of, 32
bears, 8, 41, 42
beavers, 94

tats, 37, 95–96; nonnative species in, 131; purple loosestrife as habitat, 125, 126, 127–128; vernal pools as, 57
hail, 16
Hannacroix Creek, 118
health care, 26
Heizer, Michael, 54
hemlock trees, 171, 174
hemlock woolly adelgid, 173
herbicides, 160, 161
herring, 8, 27, 115–116, 184, 193; description of, 109–110; diminishing numbers of, 106; migration of, 106, 116; spawning behavior, 107, 109; volunteer program to monitor, 106–107, 119
hibernation, 69
hickory trees, 27, 32
Hicks, Al, 31, 38
Hitler, Adolf, 130
Hollenbeck River, 66
Housatonic River, 66
Howlbox, 97
Hudson Basin River Watch, 76–77
Hudson River, 7, 9, 43, 48, 156; depth of, 79–80, 83; as eagle habitat, 189–190, 194; eels in, 12, 18, 27; emerald ash borer spread and, 171, 182; nonnative species in, 131; salinity levels, 74; salt front, 45, 49, 74, 143; tidal marshes of, 10; tides of, 107–108, 143; tributaries of, 107, 136, 138; Wappinger Creek draining into, 106; warming temperatures of, 11, 191; water chestnuts in, 45–46; water quality, 77; watershed of, 185; winter ice extent and duration, 190–192
Hudson River Almanac, 15, 17, 28, 189, 191
Hudson River National Estuarine Research Reserve, 76

Hudson Valley, 8, 12, 18, 60, 124, 144; eagles in, 184, 185; ecology database of, 16; seasons in, 109; summer in, 74, 123; warming temperatures in, 10–11
Hudsonia, 47, 124
human population, 23
humanism, 22
hunters and hunting, 12
Hunter's Brook, 106, 112, 118, 120, 193
"hurry sickness," 155–156
Huth, Paul, 3

ice pellets, 16
IceWatch USA, 204
Ifrah, George, 184
Impatiens parviflora, 130
inattention blindness, 40–41
Indiana bats, 8–9, 30, 31, 38
indigo bunting, 27
insects, 52, 57, 125, 127, 177; stewardship and, 179; timing of bird migration and, 11
Internet databases, 23
Irene, Hurricane, 3, 164, 194
irises, 55
Irwin, Alan, 19

Jackson Creek, 156, 159, 166
Japanese apricot trees, 12
Japanese beetle, 128, 163
Japanese maple, 171
Jefferson, Thomas, 22
Jefferson salamander, 57
jewelweed, 132, 157, 164
joe-pye weed, 74, 122
jumping spider, 128

Kays, Roland, 92–93, 94, 99
Kelly, Ellsworth, 81
killdeer, 11

neobiota, 133
Nest Record Card Program, 19
nettle, wild, 157
New York City, 2, 12, 20
New York Department of Environmental Conservation, 7
New York Sea Grant Extension, 76
New York State Department of Environmental Conservation, 30, 31, 86–87, 148; emerald ash borer monitored by, 169; Hudson River Research Reserve and Estuary Program, 76, 106, 136, 144, 188
Newson, Marc, 81
Niebuhr, Reinhold, 85
nine-spotted ladybug, 127
nonnative species, 6, 134; birds, 131–132, 133; insects, 170, 173; plants, 10, 123, 128–130, 133, 159, 174
Norway maple, 10

oak trees, 27, 32, 156
O'Connor, Monsignor Desmond, 162, 163
Oestrike, Rick, 159–161, 163–164, 166
Operation Migration, 19
Orth, Jennifer Forman, 127
owls, 98

pale smartweed, 157
Pascal, Blaise, 202
peach trees, 60
Penhale, Jilene and Jonathan, 126–127
peonies, 55
perch, 8, 113
peregrine falcon, 2
pesticides, 8, 50, 185
phenology, 2, 24, 108
photography, aerial, 75, 76
photosynthesis, 44, 157
phragmites, 128, 159, 207

Pickett, Steward, 140
pin oak tree, 67, 90
pine trees, 72
place, estrangement from, 13–14
plants, 127, 140; edge habitats and, 95; first bloom of, 2; flowering times of, 6, 24; nationalism and, 130; nonnative and invasive, 10, 123, 128–130, 133, 159
Plum Point, 194
Poesten Kill, 107
poison ivy, 61, 69
pollution, 20, 58, 75
pools, vernal, 7, 18, 111; cleansing function of, 57–58, 68; ephemeral nature of, 56–57, 70; as habitats, 57, 69; monitoring and regulation of, 58–59, 60; property ownership and, 66; salamander eggs in, 63–65, 68–69, 73; surveying of, 25, 62–63, 71–72
porcupines, 5
possums, 94
post offices, closure of, 13
post-normal science, 70
precision, as mathematical term, 187–188
Project Feeder Watch, 19
property ownership, 66–67
Prosek, James, 150–151
Public Laboratory for Open Technology and Science, 23–24
Public Participation in Scientific Research (PPSR), 20–21
purple finch, 27
purple loosestrife, 8, 26–27, 122–123, 132–133, 157; in autumn, 134; changing perspectives about, 124–126; as habitat, 125, 126, 127–128; as nonnative species, 123–124, 128, 133; studies of, 127

235

Index

Index